8/05

Praise for *Einstein's Heroes*:

"Arianrhod combines a passion for her subject with an erudition that is rare for a storyteller. I swear my IQ was up 20 points when I'd finished."

Robyn Williams, Presenter, *The Science Show*

"Arianrhod brings to her subject so much care, intelligence, attentiveness, enthusiasm and simmering excitement that the book reads like a good novel ... *Einstein's Heroes* is a remarkable, accessible, inspiring new book."

Jane Gleeson-White, *Sydney Morning Herald*

"A terrific book for stroppy teens demanding to know why they should learn any of this stuff, or literate adults who wish to be more numerate."

Michelle Griffin, *The Age*

"Robyn Arianrhod's passion for mathematics is so infectious, you'll scream 'Eureka' when you read her book."

Blanche Clark, *Herald–Sun*

Dr Robyn Arianrhod is a writer and mathematician whose passion for both literature and mathematics reflects her love of language. She teaches mathematics at Monash University, and her field of research includes the analysis of Einstein's equations of gravity and Maxwell's equations of electromagnetism.

EINSTEIN'S
HEROES

IMAGINING THE WORLD
THROUGH THE LANGUAGE
OF MATHEMATICS

Robyn Arianrhod

OXFORD
UNIVERSITY PRESS
2005

OXFORD
UNIVERSITY PRESS

Oxford University Press, Inc., publishes works that further
Oxford University's objective of excellence
in research, scholarship, and education.

Oxford   New York
Auckland   Cape Town   Dar es Salaam   Hong Kong   Karachi
Kuala Lumpur   Madrid   Melbourne   Mexico City   Nairobi
New Delhi   Shanghai   Taipei   Toronto

With offices in
Argentina   Austria   Brazil   Chile   Czech Republic   France   Greece
Guatemala   Hungary   Italy   Japan   Poland   Portugal   Singapore
South Korea   Switzerland   Thailand   Turkey   Ukraine   Vietnam

Copyright © 2005 by Robyn Arianrhod

First published in Australia in 2003 by the University of Queensland Press
Published by Oxford University Press, Inc.
198 Madison Avenue, New York, New York 10016

www.oup.com

Oxford is a registered trademark of Oxford University Press

Library of Congress Cataloging-in-Publication Data
Arianrhod, Robyn.
Einstein's heroes : imagining the world through the language of mathematics / Robyn
Arianrhod.
p.   cm.
Originally published: St. Lucia, Qld. : UQP, 2003.
Includes bibliographical references and index.
ISBN-13: 978-0-19-518370-2
ISBN-10: 0-19-518370-3
1. Maxwell, James Clerk, 1831–1879.   2. Einstein, Albert, 1879–1955.   3. Mathematical
physics—History.   4. Physicists—Biography.   5. Science—Popular works.   I. Title.
QC19.6.A75 2005
530.15—dc22   2004026055

9 8 7 6 5 4 3 2 1

Printed in the United States of America
on acid-free paper

For my parents,
Olwyn and Wes,
with lots of love

ACKNOWLEDGMENTS

This book had been germinating in my mind almost from the time I first fell in love with the amazing and elegant language of mathematics, but it was my agent, Jenny Darling, who finally coaxed my ideas onto these pages. She saw the project through three drafts, and I'm extremely grateful for her perceptive advice and her encouragement. I also thank Margaret Barca and Jacinta DiMase for their support.

Madonna Duffy, managing editor at University of Queensland Press, has been everything I hoped a publisher would be: innovative, supportive, enthusiastic and professional, even during the difficult period of looming deadlines and other problems inherent in the process of turning a manuscript into a book. I also thank Fiona Hobbins, Dinah Johnson, Karen Lennard and everyone at UQP for their work on the Australian edition. I am equally indebted to Simon Flynn and all at Icon Books for their belief in *Einstein's Heroes*, and their professionalism in producing the UK edition.

I also thank Bryony Cosgrove for her skill in helping me polish the manuscript; her interest in the project, and her insight and patience, have made the editing process actually enjoyable. And I thank my cousin, Associate-Professor Michael Box, and my colleague, Dr Pam Norton, for generously agreeing to read the manuscript from a technical point of view; their helpful suggestions are very much appreciated. I also thank Chris Phillips and David England for useful comments. I am very grateful to all my academic colleagues for professional and intellectual support over the years, especially Dr Norton, Dr Colin McIntosh and Dr Tony Lun.

Many thanks are due to my friend Jenny Pausacker, who encouraged me to attempt to bring the beauty and power of mathematics to a wider audience, and whose generous, ongoing advice helped me learn to write accessibly rather than academically. Her friendship has been extremely important to me, both in preparing the way for this book and in life.

Finally, I thank my beloved partner, Morgan Blackthorne, especially for endless love and nurturing, but also for invaluable practical help: hours in the library helping me with research, and countless hours reading draft after draft whilst still retaining enthusiasm for the manuscript, and always providing insightful editorial comments. Morgan, I couldn't have done this without you!

AUTHOR'S NOTE

This book invites you to explore a language I see as a celebration of the human spirit. My aim here is to give you a genuine *feeling* for this language and its history, rather than to provide an exhaustive exposition, and in the interests of brevity and readability, I have simplified the details and contexts of some of my discussions. However, I have provided sources for further reading in the Bibliography, and have added further details about some of the ideas in the Appendix and Notes and Sources. I hope you enjoy the book.

CONTENTS

A SEAMLESS INTERTWINING

In David Malouf's classic novel, *Remembering Babylon* – a rather unsettling picture of Australian colonial life – a young Englishman, Gemmy, has spent sixteen years with the Aboriginal tribe who had found him as a scrawny, illiterate thirteen-year-old, washed up on their shore, abandoned by the other sailors on his ship. On hearing that white people have settled the area, Gemmy is lured from the bush by a feeling he cannot name: the indefinable pull of racial identity, the need to belong completely to a tribe. A need fuelled by memories he can't quite grasp – and a strange conviction that he *could* grasp them if only he could remember the right words.

The settlers do want to reclaim him, although they are unnerved by his bush smell, his uncanny stealth and the preternatural feel of him. But what really frightens them is the fact that he has lost his native language – *their* language. They observe their children, chattering away in English just as Gemmy must have done long ago, and they ask themselves: 'Could you lose it? Not just language but it. It. For the fact was, when you looked at him sometimes, he was not white. His skin might be but not his features. The whole cast of his face gave him the look of one of Them.'

They consider many possible explanations for how this could have happened, but the most terrifying is that Gemmy's loss of his native language has opened him up to strange, new *thoughts*. Thoughts they cannot – dare not –

name, but which have made him part of this alien land. They conclude in bewilderment that Gemmy might have been born white, but he has become an Aborigine.

Remembering Babylon charts an interaction between language, thought, reality and identity that is fascinating in its implications, not only for how we see ourselves and each other, but also for how we see the physical world itself. In everyday situations, most of us assume that physical reality comes first, not language: we assume that we create words to describe things which already exist. We name mountains, rivers, seas, plants, animals and so on. We even create onomatopoeic words in order to describe natural sounds like the bird-songs 'cuckoo' and 'currawong', or other sounds like 'bang', 'crash', 'splash', 'thump', 'sizzle' and so on. We assume that we can use all these words objectively to describe the world around us.

However, when it comes to a more complex assessment of our surroundings, language plays a more intriguing role, and the idea that we can objectively describe physical reality becomes questionable. In speaking English, Malouf's British settlers speak a language that generally reflects their own previous experience of nature – a language which had evolved in a land of soft colours and bountiful fertility, a land long tamed by 'civilisation'. In describing their new landscape, they choose words like 'hostile', 'uninhabitable', 'empty'. They see it as something to be 'tamed', 'owned', and 'filled' (with Britishness). Gemmy sees the same land entirely differently. Far from empty, it is alive with ancestral presences, and to describe it, he would choose very different words from those used by his British friends – words with no exact English equivalent, but which would suggest interdependence rather than subjugation, bounty rather than barrenness, respect rather than fear.

For most of us, subjectivity dominates even when we try to represent our surroundings visually: one of the most important and difficult lessons art students must learn is to draw what they see, not what they *think* they see. The influence of European preconceptions about nature was highly evident in early Australian landscape painting; it took time for colonial artists to stop painting Australia as though it were Britain or Europe, and to learn to make relatively objective representations of the landscape itself. It seems that thought is inherent not only in our descriptions of reality but also in our *perception* of it; since most of us think in terms of language rather than pictures, it is not such a big step to suppose that, as Malouf puts it, 'the world as we know it is in the last resort the words through which we imagine and name it'.

Or, I would add, the equations. Physics is the science of describing the physical world at its most fundamental level, and mathematics is the language most physicists use to name and imagine that world. And I do mean 'imagine'. It is by thinking mathematically – not just by seeing and touching and doing physical experiments – that physicists have created some of their more bizarre concepts about nature. Like quantum theory, which says that at the subatomic level, energy – such as heat and light – can exist only in discrete 'packets' or 'quanta'. This is a strange concept, completely alien to our everyday experience of reality; when you boil a kettle, for example, the water temperature increases continuously from room temperature to boiling point. It does not jump straight from 20 to 25 degrees, say, then directly to 30 degrees, and so on in a series of 'quantum leaps'.

Then there's the equally counter-intuitive concept that the universe is expanding, a possibility that first turned up as part of the mathematics in Einstein's general theory of relativity. Who but a mathematician could imagine such an idea? After all, if you watch the sky night after night, year after year, the stars appear to keep to their seasonal patterns; there is no hint that they are actually moving away from each other, wrested apart by the ever-growing universe as though they were tiny specks on the surface of an expanding balloon.

Einstein's equations also suggested that the universe was actually created, in a colossal explosion we now call the Big Bang, from a single 'four-dimensional' point in 'spacetime'. A point whose apocalyptic manifestation marked the beginning of time and the creation of space itself, so that we cannot conceive of what happened before the Big Bang, because there was no pre-existing time with which to compare 'before' and 'after', and no space waiting around to be filled – by a deity or otherwise – with stars, planets and people. There was simply an incomprehensible *nothing*.

The cosmos conjured up by mathematical physicists has a host of other strange features, some of the most tantalising being that time literally slows down in certain physical situations; that 'black holes' swallow all the matter around them and cause it to disappear from the universe, perhaps squeezing it through a 'wormhole' out into a 'parallel universe'; and that 'antimatter' annihilates any nearby matter so all that is left is a burst of energy, courtesy of Einstein's famous $E = mc^2$, which can be rescaled to give $E = m$, an equation whose unearthly consequence is that ethereal, immaterial energy (E) is equivalent to – *interchangeable* with – solid matter (m).

It is the stuff of science fiction, but it is not mere fantasy: physicists can describe these scenarios in precise mathematical detail. But just as Gemmy's world is incomprehensible to the settlers who do not speak his language, and who therefore cannot think his thoughts, so is much of our universe a mystery to anyone but a mathematical physicist.

You might think that the physicists' view of the world is *arbitrarily* subjective, even irrelevant, created as it is in terms of one particular and arcane language. Indeed, some philosophers say that the ordered nature of mathematics leads physicists to impose a subjective orderliness on their notions of the physical world, just as the use of the English language imposes a degree of Englishness on a person's perception of other landscapes. However, while Gemmy and the settlers cannot agree on a verbal description of the land they share, most mathematical physicists claim that *their* descriptions of the world often turn out to have an extraordinary and unexpected degree of *objectivity*, in the sense that ultimately they can be physically demonstrated in ways we all agree upon, regardless of our culture or the other languages we speak.

For example, many people all over the world use radios, televisions or mobile phones, which work by receiving radio waves; however, radio waves were first discovered not in the physical air itself, but in a set of mathematical equations. They were mathematically imagined into being, you might say, because physicists then went looking for physical evidence of their existence. When they found it, they learned to produce these waves so successfully that no one with a television or mobile phone aerial doubts the waves exist, even if most of us do not understand what they actually are. By contrast, Malouf's settlers never manage to detect the spirits of which Gemmy is so aware. While people from different cultures can rarely perceive each other's ghosts and fairies, when physicists tell each other tales about ethereal substances dreamed up in the language of mathematics, very often they can make manifest their mathematical imaginings for all to see.

Which is not to say that other kinds of ghosts do not exist, or that this kind of objectivity is *better* than subjectivity and the rich cultural diversity it spawns. Rather, it is to delight in the deliciously uncanny fact that one particular human language (whose essence is common to all cultures, because all peoples have developed mathematics to some degree) has an unparalleled capacity to describe the physical world, including its secret, hidden parts, in prescient, objectively demonstrable ways.

For instance, most physicists now believe there is enough evidence to prove not only that radio waves exist but also that black holes abound in our universe, that time does sometimes slow down, that individual 'antiparticles' exist, and that the universe is continuing to expand from that intitial explosion. On a more direct level, the equivalence of ethereal energy and solid matter has been confirmed repeatedly in nuclear reactors (not to mention nuclear bombs), and in everyday objects like household smoke detectors and EXIT signs which glow when the lights are out (thanks to the energy given off by radioactive material inside the sign). And the physical wisdom of quantum mechanics' mindbending mathematical images of the subatomic world can be seen in the development of the microchip (to take one example), which is the basis of portable computers and a range of modern electronic devices, such as washing machines, cars and television sets.

Evidence like this was not available a century ago, and most people then thought that mathematics was as subjective as any other language – that mathematical descriptions of nature were purely metaphorical, and were therefore secondary to reality itself. Nowadays, however, many physicists believe that the universal language of mathematics is the most fundamental tool we have – and the most objective one – for exploring and describing the basic nature of the physical universe, and our place in it. They say, therefore, that mathematics is the defining language not of a tribe or even a race, but of a species.

This book is about the age-old human quest to make sense of the physical world. It is centred on the work of James Clerk Maxwell, who created the paradigm of present-day physics with his theory of electromagnetism (the theory that predicted the existence of radio waves). He was the first physicist to embrace deliberately the ambiguous relationship between language and reality; the first to accept that in a very real sense, language *is* reality. He showed that the structure of mathematical language seems to reflect hidden physical structures, so that in the unseen realms of the world, such as the heart of atoms, radio waves and black holes – 'the hidden, dimmer regions where thought weds fact', as he put it – the closest we may ever come to perceiving physical reality directly is to imagine it mathematically.

Einstein was so inspired by Maxwell that he placed a photograph of him on his study wall, and Maxwell is the hero – among many heroes – of this book, too. But while his story provides a narrative focus, this book is ultimately about *ideas*: it is a story about how physicists name and imagine the world with the language of mathematics.

A RELUCTANT REVOLUTIONARY

The past century's dazzling proofs of the awesome ability of mathematical language to uncover hidden physical secrets have moved contemporary physicists Stephen Hawking and Paul Davies to speak of uncovering the 'Mind of God'. Even in the early 1900s, Einstein had asked, 'How can it be that mathematics, being after all a product of human thought independent of experience, is so admirably appropriate to the objects of reality?' When mathematics took on its new role as discoverer of physical reality, following Maxwell's work in the mid-nineteenth century, a philosophical and scientific revolution was launched whose sophistication still amazes many physicists.

Maxwell himself, however, would have been shocked by all the fuss. He was known to his friends as a man of gentleness and humility. He also had a biting wit, with which he forthrightly attacked any tendency to self-importance in himself or his colleagues. Besides, with hindsight we have a far better appreciation of the philosophical significance of his work than he could possibly have had.

But Maxwell was not the unwitting architect of his revolutionary legacy. All physicists aim to study the physical world as objectively as possible, yet he had an unusually strong commitment to discovering the truth about Nature. To seeing it as it is, unencumbered by preconceptions. He needed an unprecedented amount of patience and insight in his search for the best way of doing this, and in the end –

despite an injunction to his mathematical colleagues not to 'flatter themselves' that their symbols enabled them to express new ideas ordinary words were as yet unable to express – he found that the use of pure mathematical language was the best way to avoid bringing everyday beliefs about reality to the creation of new physical theories.

James Clerk Maxwell was born on 13 June 1831, the only surviving child of John Clerk Maxwell and Frances Cay. John was a barrister, and the laird of Glenlair, the family estate in Dumfriesshire, south-west Scotland. The Scottish countryside filled young James with wonder as he observed it in all its seasons. He learned from the animals, too, these 'living companions of his solitude'. The frogs were his favourites; he would pick them up gently and listen to them, and he felt such kinship with them that sometimes he would put one in his mouth and let it jump out – to the astonishment of his friends and family.

John had been educated at the University of Edinburgh and as a young man, he regularly and enthusiastically attended scientific meetings of the Royal Society of Edinburgh. He was still fascinated by science and technology, and never tired of answering James's questions about how things worked. 'What's the go o' that?' young James would ask, and if not satisfied with the first answer his father gave him, 'But what's the *particular* go of it?'

Frances, too, encouraged him. She was in charge of his education, but even before he'd turned three she expressed her pride in his enquiring mind in a touching letter to her sister Jane. He was a 'very happy man', she said, and so curious about the world that ' "Show me how it doos" is never out of his mouth'. She particularly recalled his delight as he dragged his papa through the house to show him the holes in the walls where the bell-wires came through. Maxwell was amazed that when he pulled a bell-rope in one room, a bell rang in another room. He had assumed this must be due to some strange form of magic, but now he had discovered the pull was transmitted from the bellrope to the far-off bell by a hidden connecting wire. This simple but ingenious mechanical fact intrigued the young child; it was an image that would stay with him all his life, and which he would later use to illustrate his famous scientific theory.

Frances was not fated to see her son's success. She died of cancer when he was eight, and he later wrote that he felt her loss for many years, even though his father became his 'companion in all things'.

Growing up motherless, but with a father who loved him enough for two parents, Maxwell had a remarkable sense of himself. His independence was tempered in the fire of his battle with his hated home tutor, an inexperienced sixteen-year-old who had been hired after Frances' death. He was overly fond of the strap, and he would also hit his student on the head with a ruler and pull his ears till they bled. Maxwell was not impressed; his experience with his parents had led him to expect a more interactive and respectful approach to learning. But he handled the situation with characteristic self-reliance. Instead of complaining to his father, he simply refused to cooperate with the tutor, developing a strategy of never directly answering questions and of speaking hesitantly, a strategy that became a lifelong habit.

It was his Aunt Jane who, on a visit to Glenlair, realised the tutor had to go. Maxwell, then ten years old, was sent to live in Edinburgh, with his Aunt Isabella, so that he could attend the relatively progressive Edinburgh Academy. Unfortunately, he found school no less of a trauma than home tutoring. Shy, awkward and slow of speech, he wore strange clothes – made and designed by his father with comfort rather than fashion in mind – which made him such a target for the other students that he arrived home from his first day at school with his clothes in tatters.

The bullying continued, and according to his fellow student, Lewis Campbell, Maxwell generally responded to it with his 'natural weapon', irony, and did not appear to be too upset. However, Campbell described the exact moment his lifelong friendship with Maxwell was born. His classmate was again being tormented, and although 'the young Spartan' himself seemed unperturbed, Campbell felt a warm rush of chivalrous emotion which impelled him to attempt to intervene. He was rewarded by a look of 'affectionate recognition' in Maxwell's eyes, which he never forgot.

Maxwell eventually responded physically as well as verbally to the bullies, and they let him alone. After a couple of years, he settled down to his schoolwork, which he had initially found boring and irrelevant, and became one of the top students. The future mathematician, Peter Guthrie Tait, began to take notice of him then, and became another lifelong friend.

But Maxwell's best friend continued to be his father, to whom he wrote humorous and interesting letters. Some of them were decorated with drawings and addressed to his father at 'Postyknowswere' (although he did provide the posty with the appropriate region). Some of them were signed with a mix of nineteenth-century formality and relaxed, childish humour: 'Your most obedient servant, Guess who?' In one letter, he even poked fun at his own hesitant speech pattern, which his teacher was trying, unsuccessfully, to help him to correct. John hated to be apart from his son, and whenever he could afford to be away from Glenlair, he would stay with him in Edinburgh, sometimes even visiting him at school; Campbell recalled many occasions on which he saw John's 'broad, benevolent face beaming with kindness' towards his son's friends.

One of the most poignant of all John's paternal actions – and one of the most important events in setting Maxwell on the road to becoming a mathematical revolutionary – was his initiative and perseverance in contacting Professor Forbes of Edinburgh University, in order to tell him about his son's first original mathematical investigation. Maxwell was then only fourteen, and had received little formal mathematical education. His classmate Tait recalled that in Maxwell's first years at the school, before he had made many friends, he would spend his spare time 'reading old ballads, drawing curious diagrams, and making mechanical models' – activities so unintelligible to the rest of his schoolfellows ('who were then totally ignorant of mathematics') that they called him 'Dafty'. His solitary pursuits paid off, however, because he eventually produced, completely independently, a new method for constructing some unusual geometrical curves, based on a generalisation of the geometry of an ellipse (or flattened circle).

John's diary entries at this time are full of 'James's ovals'. And when Forbes said he was interested in them, and asked for something in writing, John himself formally wrote a version of his son's results (although James's own account also survives). His diary then records another ten days of his and Forbes's activity on the paper – visiting, checking, refereeing and 'cutting out pasteboard trainers for curves for

James'. Finally, after six weeks of activity on behalf of his son's 'ovals', John received the news from Forbes: 'If you wish it, I think that the simplicity and elegance of the method would entitle it to be brought before the Royal Society', which Forbes offered to do himself, since Maxwell was too young for the honour.

When the great day arrived, John proudly recorded, in his understated way, that James's paper 'met with very great attention and approbation generally'. The paper was duly published in the *Proceedings of the Edinburgh Royal Society* (6 April 1846), and in this remarkable way, Maxwell's mathematical research career began.

He soon realised, though, that he was not interested in mathematics simply for its own sake; he wanted to learn how to apply it to the natural phenomena he saw around him. But first he had to learn more about physics, the science that describes the natural world. Only then would he be able to combine his innate appreciation of mathematics with his love of nature.

BEETLES, STRINGS AND SEALING WAX

After Maxwell's paper was read at the Royal Society of Edinburgh, his father regularly took him (and often Lewis Campbell, too) to hear other lectures at the society's Monday evening meetings. Scientific lectures were a popular form of public education and entertainment, and Edinburgh attracted its share of inspiring speakers. The society met in a large, candlelit room with soft carpet and huge windows; the audience sat on cushioned benches facing a long table, around which sat the society's councillors, while the reader of the evening's paper would place his illustrative material on the table, and would read from one end of it, facing the president's chair.

It was here that Maxwell became intrigued by the physics of electricity and magnetism. Magnets – made of substances like iron, steel and a naturally occurring rock made of iron oxide (called 'lodestone' by the ancients) – have the peculiar and innate ability to attract or repel similar substances. But there is a neat, natural pattern to this phenomenon. Each magnet has a north and a south pole where the magnetism is strongest. If two magnets are aligned so that one's north pole is facing the other's south pole, they are drawn together by a magnetic force. If similar rather than opposite poles are facing each other, then the magnets repel each other. (In fact, the Earth itself is a huge magnet with north and south poles, although these do not exactly coincide with the geographical poles defined by the Earth's geometrical axis; nevertheless, the north-pointing or magnetic north alignment of a freely moving magnetic compass needle points approximately to the geographical or true north pole.)

Electricity also manifests the pattern whereby opposites attract: there are two types of electric charge – positive and negative – and two oppositely charged objects attract each other, while similar ones repel each other. The natural symmetry between two such different phenomena as electricity and magnetism reveals a precise, mathematical kind of beauty in nature, an intrinsic harmony.

However, electricity has the additional and very useful ability to 'flow' along a wire or other metal conductor. In modern terms, this means it can be transmitted along wires from electric power stations to our homes, but in the 1840s, there were no power stations and no household electrical appliances. Physicists still had no idea what electricity really was, but they did know that it could flow around an 'electric circuit'.

An example of a basic circuit is a torch, in which two wires or metal clips each have one end connected to a battery terminal, their other ends being attached to a light globe. The wires, globe and battery form a connected loop or circuit. Conventional batteries or 'cells' use chemical reactions to produce electricity (in contrast to solar cells, which use the energy of sunlight). In the simplest type of chemical cell (and others are just variations on the theme), the terminals are the external ends of metal rods, one made of zinc and one of copper, which sit in a bath of dilute acid inside the battery casing. When the terminals are connected to the wires, a chemical reaction occurs in which the acid begins to dissolve the zinc and copper, releasing in the process a flow of negatively charged particles that travels as an 'electric current' from the zinc rod to the copper one, via the wire and the globe in the circuit, and then back into the acid solution. The chemical reaction cannot continue without this exchange of electric particles, so if the circuit is broken by disconnecting either of the wires from the battery or the light globe, or simply by cutting the wire in the middle of the loop, both the current and the chemical reaction stop. If the ends of a broken wire are rejoined, the current flows again. (This is the principle of the electric switch: to turn on a torch or any electrical appliance, you press or flip a switch, which 'rejoins' the wire so the current can flow. When you release the switch, or flip it in the opposite direction, you 'cut' the wire and the current stops.)

The original electric battery was the 'Voltaic pile', invented in 1800 by the Italian experimental physicist, Count Alessandro Volta; it was a huge contraption, unlike modern batteries, most of which are single cells rather than a 'battery' or 'pile' of cells. Incandescent light globes were pioneered in the 1830s, and no doubt Maxwell saw a demonstration at the Royal Society of the astonishing phenomenon of battery-powered electric light. He would not have seen it in everyday life because light globes were not made practical until 1879, by the American inventor Thomas Edison, and contemporaneously, by the English physicist Joseph Swan. One of the first houses in the world to be equipped with electric light was the mansion built in 1881 by the famous mathematical physicist, Sir William Thomson, who later became Lord Kelvin.

The Royal Society meeting that most inspired Maxwell, however, was the one at which he saw something even more amazing than electric light: an 'electromagnetic machine'. He did not record *which* machine he had seen, but it is evident he was captivated by the new science of 'electromagnetism', in which the natural symmetry between electricity and magnetism is even more beautiful.

Electromagnetism: A new and exciting science

Traditionally, electricity and magnetism had been thought to be completely separate phenomena. By the end of the eighteenth century, scientists had begun to suspect some sort of interaction between them, although the bases of these suspicions remained *ad hoc* until 1820. In that year Danish professor Hans Oersted formally announced a remarkable discovery. He had noticed that if a battery was connected to a loop of wire so that an electric current began to flow through the wire, a nearby magnetic compass needle would jump, as if attracted by another magnet – as if the electric current itself were a magnet! Oersted came to the astounding conclusion that an electric current creates (or 'induces') its own magnetic force, apparently out of thin air because there is no force coming from the battery or the wire when the current is not flowing, and there is no magnetic force from a stationary charged object (like an electrified ball made of non-magnetic metal). It is only when electric charges are *moving*, as in a current, that a magnetic force appears.

Although they had no idea *how* an electric current created a magnetic force, scientists were quick to make use of this newly discovered gift from nature. One of its most important applications was the galvanometer, an instrument for measuring electric current. (After Oersted's momentous discovery, it was soon realised that the number of degrees through which a compass needle is deflected by an electrically produced magnetic force gives a measure of the strength of the electric current that produced it.) However, the most dramatic application was the telegraph.

By the time Maxwell saw his first 'electromagnetic machine' in the mid-1840s, telegraphy was just beginning to revolutionise nineteenth-century communications. The discovery of the amazing link between electricity and magnetism suggested the marvellous possibility of 'instant' communication, anywhere in the world, because electricity seemed to flow from one place to another almost instantaneously. (In practice, though, currents could not yet travel very far without overheating the wires or otherwise being impeded or distorted.)

As first shown in 1821 by the great French physicist, André Marie Ampère, telegraphic messages can be sent by switching an electric current on and off; the intermittent current travels down a wire to the receiving end, where a compass needle or other magnet acts as a 'receiver'. The bursts of current automatically induce bursts of magnetic force which magnetically attract and release the 'receiver' magnet every time the current is switched on and off. Ampère's receiver was a magnetic needle mounted on a dial on which letters were painted; each time the current was switched on, the deflected needle pointed to a letter. (You could choose which letter you wanted the needle to point to by regulating the amount of current used.)

Inspired by Ampère's prototype, physicists and inventors began working on a practical needle telegraph system. These pioneers included the Russian, Pavel Schilling, the German, Karl Steinheil, and the Englishmen, Charles Wheatstone and William Cooke. A complex five-needle system developed in 1837 by Wheatstone and Cooke was being used to coordinate the new and increasing steam-train traffic between local British railway stations, but most of the early inventors found it hard to convince governments and investors that telegraphy had any practical use.

Even the physicists tended to be unaware of (or uninterested in) any practical applications of their electromagnetic researches, although many of them contributed to the further development and improvement of the telegraph during the 1830s – notably the American, Joseph Henry, and the Germans, Carl Gauss and Wilhelm Weber. It was the American inventor, Samuel Morse, who combined their results into a practical, *long-distance* telegraphic system. Well aware of its commercial potential, he took out a number of patents in the 1830s and 1840s, and spent much time trying to raise government funding to implement his inventions.

Building on the work of Henry, in particular, Morse had developed a powerful transmission system which enabled electric signals to travel over longer distances than had been achieved previously, and in 1844, government funding having been secured at last, he sent the world's first long-distance telegraphic message over the 65 kilometres between Baltimore and Washington.

His most famous telegraphic system – developed in the late 1840s, using the scientific discoveries of Henry, Steinheil and William Sturgeon – used as its receiver a spring-loaded magnet to which was attached a pencil. The set-up was such that when the current was turned on, it produced a magnetic force that pulled the pencil-holding magnet towards a continually moving strip of paper so that the pencil touched the paper and left a mark; as soon as the current was switched off, the pencil was released. If the current was on for a very short time, the pencil only left a tiny dot of a mark, but if the switch (or operating key) was held down for a longer time, the induced magnetic force kept the pencil on the moving paper for long enough to produce a dash rather than a dot. Messages could then be sent in 'Morse code', a system in which the letters of the alphabet were coded in terms of combinations of dots and dashes (the famous international distress call SOS being . . . - - - . . .). Interestingly, it was soon realised that the code could actually be *heard*, as a series of taps separated by longer and shorter intervals, depending on whether the pencil was recording a dash or a dot, and so a proper sound receiver was developed. (Coded telegraph messages are still used today, in the fax and telex systems.)

Morse's achievements later became clouded by his refusal (for fear of losing his patents) to acknowledge the extent to which he had built on the experimental and theoretical work of Henry and others, but in 1846, telegraphy would still have been big news at Royal Society and other scientific meetings around the world; Morse's 1844 triumph had inspired a frenzy of telegraphic cable-laying in the United States, Europe and Britain, where the first submarine cable had just been laid in Portsmouth Harbour. Perhaps some sort of telegraphic apparatus was the electromagnetic machine Maxwell saw demonstrated in Edinburgh.

Or perhaps he saw an early version of the electric motor, a prototype of which had been first designed 20 years earlier by the extraordinary English physicist, Michael Faraday. A motor is a machine that moves things – the wheels of a train, or the moving parts of a computer, a washing machine, an electric fan, a power drill, a food processor or a CD player. All these and many other modern household and industrial devices are run by electric motors. The original motor was the more cumbersome steam engine, whose rotor – the part of the motor which is attached to the wheel or other part which needs to be moved – was turned by the energy of steam produced from water heated by burning coal or other fuel. Faraday built the prototype of the electric motor in 1821, using the fact that a magnet actually rotates when it moves under an electrically induced magnetic force. (For example, the deflection of a compass needle is a partial rotation. Continual, complete rotations can be achieved by appropriately regulating the amount and direction of the current.) If a magnet is attached to a rotor, then when the magnet rotates under the influence of the magnetic force created by a nearby current-carrying wire, the rotor rotates along with it. Practical electrical motors were developed over the decades, by Henry among many others, and the first large-scale use of these motors was in the electric trams of the 1880s. A prototype or toy version of such a motor would have been a novelty in the 1840s when Maxwell was a schoolboy.

Faraday's technological genius had not stopped with pioneering the electric motor; he figured that if an electric current – that is, a flow of moving electric charge – could produce a magnetic force, then pehaps there was a natural symmetry so that a moving magnet could produce an electric current in a loop of wire which was not attached to a battery or other source of electricity. He was right, as was Joseph Henry, who independently made the same discovery at around the same time. Faraday used one of the newly invented galvanometers, attached to the loop, to confirm that an electric current really had been created.

With this discovery, Faraday built a prototype of the electricity generator, the exact reverse of the electric motor.

In the generator, a rotor is attached to a magnet as before, but it has to be mechanically turned in order to move the magnet, which then induces an electric current in a surrounding loop of wire. The rotors in modern commercial generators are turned using the power of running water (as in hydro-electricity), of wind, or, most commonly, of steam. As in the early steam engines, it has been traditional to produce the steam for generators from water heated by burning coal; nowadays, it can also be produced with the heat released during nuclear fission (nuclear energy) or heat from the Sun (solar energy). Because of the seemingly magical ability of a moving magnet to create electricity out of thin air, solar- and wind-powered generators can deliver cost-free, pollution-free electricity, apart from setting-up and maintenance costs (although wind generators are currently problematic for aesthetic reasons).

Like Henry and other physicists who were interested in ideas rather than things, Faraday had not seen any commercial use for his inventions, regarding them primarily as tools for demonstrating electromagnetic principles. It was the exotically named Frenchman, Hippolyte Pixii, who in 1832 turned Faraday's prototype generator into a workable machine. Today, commercial generators send electricity along wires to towns and cities, but their first large-scale use was to supply electricity for lighthouses and factory lighting in the 1870s. Even in the 1830s and 1840s, however, Faraday's discovery was being applied: Cooke and Wheatstone used a generator to ensure a reliable source of electricity for the telegraphic messages they sent between railway stations.

The natural phenomenon whereby electricity is generated by moving a magnet is called electromagnetic induction; it had been discovered by Faraday in 1831, the year Maxwell was born. Together with Oersted's discovery, it had launched the modern electricity-based technological revolution. These two discoveries also showed that electricity and magnetism are inextricably linked, each one capable of inducing the existence of the other in a phenomenon called electromagnetism. It was dramatic new science, and yet its principles could be demonstrated to the public with such simple equipment – a coil of wire, a magnet or two, a battery, a galvanometer – that young Maxwell was inspired to set up his own laboratory in an attic at Glenlair.

He used an old door set on two barrels for a table, and his experimental tools consisted of a few jam jars, batteries, pieces of wire, and some bar magnets given to him by his father. He set himself the task of exploring the conduction of electricity, which was still not fully understood. It would not be explained for another 50 years, when the English physicist, J.J. Thomson, discovered the existence of electrons, the negatively charged sub-atomic particles that flow from atom to atom in materials (like metal wires) which are good electrical conductors. They are the same charged particles, we now know, which are released from the atoms in the zinc as it dissolves in the acid inside a battery, and which then flow as an electric current along the wire in the circuit. In Maxwell's day, however, many scientists thought electricity was a kind of fluid – hence the electrical application of the terms 'current' and 'flow' – and even those who tentatively spoke of particles or molecules of electricity had no idea what these 'molecules' might be, or how they behaved.

One of Maxwell's most innovative experiments involved testing whether a beetle might be a good electrical conductor (it wasn't!) but he noted in a letter to Campbell that his experiment was 'not at all cruel', because he had made sure the beetle was dead before he electrocuted it, giving it a painless death by drowning it 'in boiling water in which he never kicked'.

Newton's theory of gravity: Still producing exciting new discoveries

At another Royal Society lecture, Maxwell became inspired about a different area of science – one in which mathematics and the study of nature had been beautifully integrated. In 1687, the famous Englishman, Sir Isaac Newton, had published the first self-consistent, accurately predictive theory of physics: the mathematical theory of gravity. Gravity itself was a well-known phenomenon: it is the natural force that pulls everything towards the ground. But Newton was able to formulate a mathematical law – an equation – that accurately described the effects of this ubiquitous but little understood downward pull. For instance, if you were to throw a ball into the air with a certain speed, Newton's mathematics could tell you beforehand how high the ball would reach, and how long it would take to fall back to the ground. The most astounding part of Newton's theory, however, was its use of mathematics to prove a brilliant and radical hypothesis that laid the foundations of the modern science of astronomy: the hypothesis that gravity is a universal force, not just an earthly one as everyone had supposed, and that the ancient mystery of why the planets move through the sky can be explained by assuming the Sun's gravity is responsible for holding them in their orbits.

In the 1840s, Newton's theory of universal gravity was still the only comprehensive physical theory known; an enormous amount of information about other topics (like electricity and magnetism) had been built up over the centuries, but no satisfactory explanatory theory had yet emerged in any of these areas. In particular, no complete *mathematical* formulation had been made in any science other than astronomy. The advantage of being able to express a theory or hypothesis mathematically is that testable physical predictions can be made – like how long the ball will take to fall to the ground or how long a comet will take to reappear in the sky. The high success rate of various tests of Newton's theory over the previous 160 years had made astronomy the most glamorous and prestigious of the sciences. The audience at the Royal Society meeting that Monday night in early 1847 was abuzz with excitement about the latest Newtonian success: the amazing story of the mathematically based discovery of a new planet.

The story had begun with the fact that Uranus, the outermost planet known in the early 1840s, had an orbit that was *not* predicted by Newton's theory ... unless there existed another planet, as yet unseen, whose own gravity was large enough to distort Uranus's orbit. (You can see how gravity distorts the paths of moving objects by hitting a tennis ball horizontally through the air; its path will not remain horizontal for long, but will be distorted by the force of the Earth's gravity so that it arcs downwards towards the ground.) Newton's mathematics described the curved path each planet should take in its journey around the Sun, but Uranus's orbit was distorted even more than expected, according to calculations of the combined gravitational forces exerted on it by the Sun and the known planets.

Apparently inspired by the famous Scottish science writer, Mary Somerville, who had publicly suggested the explanation that another planet could be affecting Uranus, the English astronomer, John Couch Adams, worked through the mathematics and eventually showed that Newton's theory *would* account for the observed distortion of Uranus's orbit if there existed another large planet. Unfortunately, the British astronomer royal, Sir George Airy, did not initially show any interest in Adams's prediction, because at that time, he was one of the few who doubted the exactness of Newton's law. Meanwhile, in France, Urbain Leverrier had been working independently on similar calculations, and he managed to persuade the Berlin Observatory to look for the predicted planet in the place in the sky where he had calculated it should be. The Berlin astronomers found it, in 1846.

The discovery of Neptune was said to be a 'sublime' demonstration of the power of mathematics, and no doubt such talk excited and influenced the teenaged Maxwell; he and Tait became interested in the mathematics of Saturn's rings, and ten years later, Maxwell would become famous among British scientists by winning Cambridge's Adams Prize for his paper on the topic.

(Saturn's rings had been assumed to be solid, as they appear to be when seen through a telescope. But by the 1850s, mathematicians had begun to wonder whether they might be fluid, because Newton's laws predicted that solid rings would fragment under the force of gravity. For his prize-winning essay, Maxwell carried out the complex and laborious mathematics which showed that the rings would be stable under Newton's theory if they were neither fluid nor solid, but were made up of many smaller solid bodies. Airy called Maxwell's paper 'one of the most remarkable applications of mathematics to physics that I have ever seen'. His judgment proved well founded 125 years later: in the 1980s, the United States space probe *Voyager* found that Saturn's rings are actually hundreds of ringlets composed of myriad particles, probably made of ice. Another sublime triumph for Newton's mathematics, and for Maxwell.)

Leaving school

At school, Maxwell, Tait and Campbell shared the top mathematical honours; the other students were far behind them. In late 1847, at sixteen, they left school and went on to the University of Edinburgh, where they became Professor Forbes's students. Campbell and Tait transferred to English universities after a year, Campbell going to Oxford (after which he became a classicist and an Anglican minister) and Tait to Cambridge (where he began his career as an academic, lecturing and researching in mathematics and physics). Staying behind at Edinburgh – probably to please his father, who would have missed him had he gone further afield – Maxwell began reading Newton and trying to rediscover the great man's work for himself: 'There is no time of reading a book better than when you ... are on the point of finding it out yourself if you were able', he wrote to Campbell.

After completing his three years at Edinburgh, Maxwell decided to go to Cambridge. He was still considered eccentric (if no longer 'daft'), and he does seem to have been a strange dinner guest, answering questions indirectly and in a monotone, if he heard them at all. He was more than likely absorbed in observing the refraction of the light in his glass of water, or some other such natural interplay. But not everyone was fazed by his odd manners; Mrs Campbell, for one, always made him feel at home. He wrote to Campbell in later life, 'I shall always remember your mother's kindness to me, beginning more than twenty-three years ago, and how she made me the same as you [and your brother]'. For her part, she had seen him as 'full of genius', and she thought Cambridge would rub the odd edges off and he would become a 'distinguished man'.

His mentor, Professor Forbes, thought so, too. Writing a series of letters to introduce his protégé to the master of Trinity College at Cambridge, Forbes said,

> 'He is not a little uncouth in manners, but withal one of the most original young men I have ever met ... He is a singular lad, and shy, [but] very clever and persevering ...
>
> I am aware of his exceeding uncouthness, as well mathematical as in other respects ... I thought the Society and Drill of Cambridge the only chance of taming him and much advised his going ... I should think he might be a [scientific] discoverer ...'

Like many Scottish Presbyterian parents of the time, John Maxwell was worried about England's potential social effects on his upright and religious son, but in the end, he gave his complete blessing and young Maxwell left Scotland for Cambridge, to become a physicist like Newton.

In order to follow the rest of Maxwell's story, however, it is necessary to take a detour, to meet Newton himself, along with some of the other founders of modern physics. And to further explore just what physics is, and how mathematics comes into it.

THE NATURE OF PHYSICS

As Maxwell would later tell his students, physics tells us absolutely nothing about emotions, spirituality and other 'higher' aspects of our experience of life. Instead, it is concerned with the fundamental structure of the natural world – the way it is put together, and the reasons it works the way it does. (Why the planets move, for example, or how and why electromagnetic induction happens.) Physicists study the basic building blocks of the physical world, like matter, gravity and electromagnetism, rather than its more specialised aspects, such as the nature of the Earth and its constituent elements, which are the subjects of geology and chemistry, respectively, or life itself, which is the subject of biology.

Consequently, physics is considered to be 'simple' or 'basic', in terms of the physical processes it studies. However, as Maxwell would also say, this very simplicity is the key to physics's predictive power, because it enables it to be 'better understood and more perfectly expressed than those [disciplines] which relate to higher things'. The language that enables this 'perfect expression' is, of course, mathematics.

The fundamental grammar of the mathematical language is based on the simple rules of arithmetic, which define the relationships between *numbers*: addition and subtraction, multiplication and division. But these rules can be generalised and extended into the fascinating language of algebra. Algebra uses letters as well as numbers; it is the language of mathematical equations, like $E = mc^2$, in which letters represent *concepts* like energy and mass, which can ultimately be quantified and expressed *in terms of numbers*. The use of letters to represent arbitrary or unknown numbers or numerical concepts was employed by the Egyptians and Babylonians 4000 years ago; significant advances in the subject were made in the seventh to twelfth centuries by the Indians – then known as the Hindus – and the Islamic Arabs, the contribution of the latter having been immortalised in the very name with which mathematicians describe this branch of mathematics: the word 'algebra' comes from the Arabic 'al-jabr'.

The natural world studied by physicists is full of quantifiable concepts. Not just energy and mass, but even more basic things like time and distance. Time passes from one day to the next, one season to another, and distance (as well as time) is embodied in the Sun's daily journey across the sky, or in the time it takes to walk from one place to another. Time and space are fundamental, interrelated ideas that have always been intrinsic to human notions of physical reality, and they are both expressible in terms of numbers – numbers of days, hours, seconds, for example, and numbers of steps, feet, metres.

Even the seasons themselves, so beloved by poets, so rich in metaphorical imagery, can be defined quantitatively. Summer and winter are marked by the solstices, the two days of the year when the Earth's position on its annual orbit is such that the Sun appears to rise to its highest and lowest distances above the horizon. Spring and autumn are defined by the equinoxes, when day and night are theoretically of equal duration. These are ancient definitions, apparently used in designing megalithic monuments like Stonehenge and New-grange, which measure and mark these significant days.

You can also mathematically describe how hot or cold it is, in numbers of degrees. Or the various colours of light and the sounds of music, in terms of their wavelengths or their frequencies: each wavelength or frequency has a different numerical value, and produces a sound or colour of a different pitch or hue. You can also numerically describe the strength and direction of forces like gravity, or of winds, electric currents, sea and air currents, heat flow; or the volume of rainfall; the height of mountains; the density and weight of matter. All these and countless other natural things can be quantified in some way, and can therefore be described in the language of mathematics.

On the face of it, reducing nature to numbers seems a mundane if not downright insensitive thing to do, but that is how the mathematical magic comes into it: these numbers often reveal interesting natural *relationships* or *patterns*, which can be generalised into algebra and used to describe the way nature works, or to make predictions about it in the form of equations (like Newton's equation of gravity, or $E = mc^2$); in this way, the simple rules of arithmetic metamorphose into the powerful language of physics.

There are actually two kinds of physics: experimental and theoretical. Experimental physics is the sophisticated development of our capacity to learn things about the world by observing it in its totality, and also by touching and tasting it, and pulling it apart – the way children do when allowed to exhibit their natural curiosity. We are all born experimental physicists, you might say, although 'real' physicists try to quantify their experimental results so they can express them mathematically. Theoretical physicists hypothesise about *why* things are as they are, and, following Newton's lead, they also express their ideas in terms of mathematical language. Einstein summed it up by saying there is an innate, human 'passion for comprehension, [which] is rather common in children, but gets lost in most people later on. Without this passion, there would be neither mathematics nor natural science.'

Indeed, people have been experimenting and hypothesising for millennia, but the first systematic, non-mystical approach to understanding the world seems to have developed in ancient Greece, from the sixth century BC, as the discipline of philosophy, a method of enquiring into the nature of physical reality using reason and argument. At this time, the Greeks also began to develop systematic forms of mathematics, and the two disciplines laid the intellectual foundations for modern physics. Other contemporaneous systems of thought, particularly Hinduism as recorded in the Upanishads, and Chinese Daoism, also contained elements of organised philosophical enquiry, and other ancient peoples also produced relatively highly developed mathematics; but for historical and philosophical reasons it was the Greek traditions that most influenced the development of modern science.

One of the greatest of the ancient Greek philosophers was Aristotle. He wrote perceptively on ethics, politics, psychology, logic, language and science – including physics. However, while modern physics is based on quantitative experimental observations of the physical world, Greek (and other ancient) philosophy was based on *intuitive* arguments about it. For example, Aristotle was concerned with such things as how to identify the difference between natural and artificial objects (he defined natural ones as having an 'innate impulse to change', which crafted objects, like beds and cloaks, do not have, except insofar as they are made of natural materials). He also discussed the relationships between matter and form, cause and chance, teleology and necessity, and how these concepts apply to Nature, as opposed to human endeavours like craft and mathematics.

Modern physicists embrace Aristotle's systematic approach to understanding nature, but they are not concerned with philosophical questions about a natural object's existence or purpose; they are interested in its experimentally observed physical characteristics: whether it is solid and rigid, or whether it flows (like a fluid); whether or not it is able to produce electricity or magnetism; its weight, density and other measurable properties; the way it moves in various situations – its speed and direction, for instance, or the time it takes to fall a certain distance. And while philosophy traditionally lacked an experimental basis, it was different from modern theoretical physics, too, because philosophical arguments about reality were based on logical deductions made with ordinary language, whereas theoretical physicists make deductions on the basis of mathematical language rather than everyday grammar, and on the basis of experimental physicists' observations rather than everyday intuition.

Of course, the ancient Greeks didn't just philosophise – they also had mathematicians, theoreticians and practical observers of nature, especially in the field of astronomy. Aristarchus and Hipparchus are but two of the most famous Greek astronomers, while in the field of mathematics, Pythagoras is still a well-known name among high school students; apart from his work in pure mathematics, Pythagoras and his followers built vital foundations for modern physics, particularly with their discovery of the mathematical nature of music, although much of their application of mathematics to the physical world tended to be mystical rather than what we would now call scientific. Archimedes, on the other hand, was a thoroughly modern physicist. He was also one of the best mathematicians the world has known. Born in Syracuse, Sicily, around 287 BC – nearly 300 years after Pythagoras, and a century after Aristotle – he is famous for leaping out of the public baths and running home naked, shouting 'Eureka!' (I've got it!) because he had just discovered the fundamental principle of hydrostatics (which is extremely useful for designing boats and other floating objects). Observing the rising level of his bath water as he stepped into it, he had a flash of insight which ultimately led him to explain the ability to float in terms of a numerical relationship between the weight of the water the floating object displaces on immersion and the force holding it up and keeping it buoyant.

Other ancient cultures produced experimental and mathematical scientists, too, particularly in the area of astronomy. After the beginning of the second millennium, there arose an increasing number of 'modern' scientists, like the fourteenth-century Frenchman, Nicole Oresme, who anticipated Newton in his detailed mathematical study of motion – a study of the relationships between the distance covered, the time taken and the speed travelled in a journey.

As far as Western historians know, however, in all early cultures this kind of 'modern' physics, in which physical observations are quantified and expressed mathematically, was more individual, less systematic, than the study of philosophy – or of mathematics itself. For example, the thirteenth-century Chinese mathematician, Yang Hui, had also discussed the mathematics of motion, but the purpose of his study seems to have been the illustration of a mathematical method rather than the analysis of a physical phenomenon. The birth of physics as a mathematically and experimentally based *discipline* is generally considered to have taken place in seventeenth-century Europe.

One important pioneer of the new discipline was the English philosopher, Francis Bacon, who argued against the use of Aristotle's method of logic as a means of understanding nature, suggesting that experiment is the best way of gaining such knowledge. But Bacon did not make any significant experimental discoveries himself. Some say the modern era of physics began with the work of Bacon's older contemporary and countryman, William Gilbert (who was also Queen Elizabeth I's doctor). His book on magnetism, *De Magnete,* was published in 1600, and it contained the results of his experimental studies on magnetism and static electricity, including the discovery that the Earth itself behaves like a magnet. Gilbert also developed a method for producing magnets out of suitable materials, and he was the first to use the terms 'magnetic pole' and 'electricity'.

The most famous of the early founders of modern experimental physics, however, is Bacon's Italian contemporary, Galileo Galilei. He was born in 1564 and died in 1642, the same year his successor Newton was born. Galileo was a flamboyant character whose story has taken on legendary proportions. It is true that some of the ideas popularly attributed to him were known to others, but his own experiments were so ingenious, and he wrote up his results in such an original and comprehensive way, that he is often considered to be the father of modern science. In intellectual disciplines like science and philosophy, it is often not so much priority of discovery that counts, but intelligibility. Aristotle is claimed as a founder of philosophy because in each of his books he organised a variety of disparate ideas – some of them not his own – into an orderly, original system of thought. So it was with Galileo – and with all the great historical figures of science, to a greater or lesser extent. The incomparable Newton himself said, 'If I have seen far, it is because I have stood on the shoulders of giants'.

Experimental physics

The nature of experimental physics is illustrated by a famous experiment conducted by Galileo. He wanted to find the answer to the question, 'Does a heavy object fall to the ground more quickly than a light one?' People had long assumed that it does, because a feather floats to the ground slowly while a cannon ball drops straight down. It seemed obvious that gravity must pull heavier objects to the ground faster than it pulls lighter ones. But, like the unsung heroes before him, Galileo was not content with the obvious; he wanted to know whether this difference in falling rate was due to the natural, downward, gravitational pull itself, or whether it was due to the different air resistances experienced by the different objects.

With such extraordinary care that his grand reputation seems well justified, he set about finding a definitive answer – although he did not actually drop things from the Leaning Tower of Pisa as the myth suggests. He minimised the air resistance by rolling balls down a highly polished, gently inclined plane rather than dropping them from a great height (in either case, the downward motion is due to gravity). After more than one hundred experiments, he concluded that when he rolled the balls from the same height, the heavier ones and the lighter ones all fell at the *same* rate. You can test this yourself by dropping a pen and a sheet of paper: the pen hits the ground first. But if you repeat the experiment with a tightly *crumpled* sheet of paper – which, like the pen, has much less air resistance than the flat sheet – then both the pen and the paper land at the same time, as far as you can tell by eye without making precise measurements. Galileo's result was the first piece in the puzzle of the strange phenomenon of gravity: whatever it is, it does not cause heavier things to fall faster than lighter ones.

It would take Newton to show the full significance of this fact, but meantime, Galileo's experiment showed something definitive not only about gravity and its effect on falling objects, but also about the emerging discipline of physics itself: it is the task of physicists, as well as lawyers, to carefully obtain all the facts before they draw any conclusions. Things fall to the ground because of gravity: that is one fact. A feather floats to the ground more slowly than a cannon ball: that is another fact. Putting these two facts together, it is reasonable to conclude that gravity causes heavier things to fall faster than lighter things; no less a philosopher than Aristotle had thought so. But another relevant fact is that the shape of a feather is particularly susceptible to air resistance. When Galileo tested the effect of gravity alone, without the extraneous influence of air resistance, he found it acted differently from what most people had intuitively thought. His experiment sounded a warning to philosophers and priests: in order to obtain reliable information about the physical world, the traditional philosopher's tools of intuition and logical argument, along with the priest's faith, must be combined with the physicist's detailed experimental analysis.

Properly designed experiments are so important in physics that whenever someone makes an experimental observation like Galileo's, scientific methodology requires others to do independent experiments to make sure the original experiment was accurate, and to check the experimenter's interpretation of the results. Modern experimenters have confirmed Galileo's results, using his method, and with space-age advantage NASA astronauts also repeated Galileo's experiment; they performed it on the Moon, which has no air, and so provides a laboratory with no air resistance. They dropped a hammer and a feather and confirmed that they fell to the ground together.

Galileo's careful experiments marked the beginning of the end of traditional philosophy's dominance of human attempts to understand physical reality. Since then, physics has replaced philosophy as the primary method of gaining fundamental knowledge about physical reality. (Contemporary philosophy focuses instead on issues like ethics, epistemology, language and meaning, although it still has important things to say about the physical world, particularly about the relationship between language, perception and reality; indeed, Maxwell's prowess as a philosopher is important in this story.)

Galileo's work also marked the beginning of the rise of mathematics as the language of physics; Galileo himself believed 'the book of nature' is written in this language, and, like most modern experimental physicists (and a few earlier ones), he attempted to express his experimental observations and hypotheses quantitatively.

For instance, in order to conclude that, under the influence of gravity alone, all objects fall at the same rate, Galileo first had to define what he meant by 'rate'. Before his time, when it came to thinking about how things move, many philosophers thought primarily in terms of rest or motion: objects were either still or they were moving. After all, in those days, people walked or rode a horse if they wanted to get somewhere, and the subtleties of motion were less obvious than they are to today's car drivers or train travellers. But like Oresme and his Oxford predecessors three centuries earlier, Galileo realised that motion includes not only the concept of speed but also that of *acceleration*, the *rate* at which the speed *changes*. Newton would give this idea its ultimate mathematical definition when he invented calculus, but in terms of everyday language, acceleration is the amount of increase (or decrease) in an object's speed in a given time interval. This means, for example, that it is a measure of how quickly you can get your car up to the speed limit when you put your foot down (on the 'accelerator') or how quickly you can slow down when you put your foot on the brake.

Speed itself is the amount of *distance* travelled in a given time, and is measured in kilometres per hour or metres per second (or a ratio of some other distance and time units, although metres and seconds are the standard units generally used in physics). Acceleration is the change not in distance per time but in *speed* per time, so it is measured in units like 'kilometres per hour per second', or 'metres per second per second'.

For instance, if you accelerate smoothly from rest to 60 kilometres per hour in 5 seconds, your speed increases by 12 kilometres per hour each second for 5 seconds, until you reach a speed of 12 x 5 = 60 kilometres per hour. The fact that the increase in speed is the same each second (in this particular example) is expressed mathematically by saying it is *constant* – that is, unchanging: it is 12 kilometres per hour per second, which can be more compactly written as 12 km/hr/sec. The 'per' and the '/' mean 'divided by', so this can be converted to standard units by converting the kilometres to metres and the hour to seconds, so that 12 km/hr/sec = 12000 metres/3600 sec/sec = 3.33 ... metres/sec/sec. The dots indicate that this is an 'infinitely repeating' decimal, with an infinite number of 3s after the decimal point. As an ordinary fraction, it is just the number $3^1/_3$ and in decimal form it can be written approximately as 3.33.

In an era before accurate clocks had been invented, let alone stopwatches or speedometers, Galileo measured time intervals by laboriously measuring the weight of water flowing out of a tank during the period in which each ball was rolling towards the ground, and on the basis of such painstaking measurements, he concluded that the acceleration due to gravity of an object falling from a given height is also constant, regardless of the object's size. Hence his famous counter-intuitive conclusion that, under the influence of gravity alone, and considering the fall to begin at relatively small, 'everyday' distances above the ground, all objects fall at the same, constant rate – which turns out to be 9.8 metres/sec/sec. In other words, the effect of gravity on a falling object is similar to the effect on a car (and its occupants) of putting your foot down on the accelerator: when you drop something, it picks up speed as it falls, at the rate of 9.8 metres per second every second, just as your car picked up 12 kilometres per hour (or 3.33 metres per second) every second while you were holding your foot hard on the accelerator.

Others had wondered whether gravity had the same effect on the motion of all objects, but Galileo's result was compelling in its precision. It is experimental and mathematical precision that differentiates physics from all other forms of enquiry into the nature of physical reality.

THE LANGUAGE OF PHYSICS

Galileo had answered definitively an age-old question about the effect of gravity on falling objects, and he made a detailed experimental and mathematical analysis of motion in general. But it was Newton who systematically explored the *cause* of motion. He supposed that things only begin to move when they are forced to; not by the will of God, as many earlier philosophers had supposed, but by the application of a physical *force*, like a push or pull. He then asked himself questions similar to these: Are there any forces acting on a teapot sitting still on a table? What about on a billiard ball rolling smoothly along on a highly polished table?

For most of us, the intuitive responses to these questions would be 'No' to the first, because the teapot isn't moving, and 'Yes' to the second, because the ball *is* moving. That is what most of Newton's predecessors thought, too. They assumed that objects were either still or moving, and that if they were moving, a force – or a divine impetus – must be causing their motion, and if they were still, then no force or impetus was acting.

Newton concluded that the relationship is not quite so simple as that. A push or pull does not cause motion itself, he said, but a *change* in the motion – an acceleration. The fact that the teapot is not moving does not *necessarily* mean there are no forces on it, and the fact that the billiard ball *is* moving does not necessarily mean there is a force causing its motion. Rather, you need a force to get an object moving in the first place – to accelerate it from rest to motion – but you do not need to apply any force to *keep* it moving at the same speed. It is all about using language precisely.

Take the billiard ball: if you hit the ball across the polished table, once you have started it moving, and you do need sufficient force from the billiard cue to get it rolling, it will keep going without the need for any *continued* force from the cue to keep up its motion. On a completely smooth (frictionless) table – on the Moon where there is no air resistance, either – the ball would keep moving at the same speed forever if the table were long enough. No such perfectly smooth table exists in the physical world, and so the ball will eventually slow down because even a highly polished table exerts some friction on the ball – friction being another force, which acts to *change* the ball's motion, just as air resistance slows down the falling feather (and, to a lesser extent, the billiard ball).

In the mid-1660s, Newton codified these insights into what are now known as Newton's First and Second Laws of Motion. The First Law says, in ordinary language, that if no forces are applied, then an object's speed and direction do not change: if it is not moving, its speed remains zero, and if it *is* moving at a given speed in a particular direction, it continues moving at that speed and in that direction. If you want to speed it up or slow it down, or change its direction, or start it moving in the first place, you have to force it to accelerate. (The French philosopher–mathematician, René Descartes, had come up with a similar idea, but for philosophical rather than physical reasons. Both Galileo and the Dutch physicist Christiaan Huygens had made a more scientific statement of this law, but it seems Newton was only indirectly aware of their work.)

Newton's Second Law expresses the *quantitative* relationship between the amount of force applied and the resultant acceleration. To get an idea of how this relationship was worked out, think about pushing heavy furniture around. The harder you push, the more acceleration you produce (that is, you get the furniture moving sooner and faster in a given time). But the heavier or more massive the piece of furniture, the harder it is to accelerate, so you need to push even harder to get it moving. These are intuitive ideas, gained from ordinary physical experience. To make them more precise – to quantify them – you would have to do experiments.

You would find that if you doubled the amount of force – by getting two equally strong people to push – then you would exactly double the amount of acceleration achieved. In other words, the amount of acceleration produced when you push a heavy object is directly proportional to the strength (or force) of your push. Proportionality means that the force and the acceleration are related to each other but are not necessarily equal to each other. You would write this mathematically as $F = k \times a$, where 'F' is the strength of the force of your push, 'x' is the multiplication sign, 'a' is the measured acceleration, and 'k' is an unknown 'x-factor' relating 'F' and 'a'. The proportionality works mathematically like this. If 'k' happened to be 3, for example, so that $F = 3 \times a$, and if you then double the acceleration or 'a' – that is, you multiply it by 2 – you get a new force whose strength is equal to $3 \times (2 \times a)$, or $6 \times a$; this force has twice the strength of the original one, which was only $3 \times a$, so by doubling the amount of acceleration in the equation, you automatically double the amount of force. This is what you would expect physically: if you want to produce twice as much acceleration – to get the furniture increasing its speed twice as quickly – you *have* to double the force. Similarly, if you double 'F' first, you mathematically *have* to double 'a' to keep the equation balanced. If F = 3 x a, then 2 x F = 3 x (2 x a); you cannot double the 3 factor, because it is fixed for this particular situation. If you push twice as hard, you *automatically* get twice the acceleration.

You can see from this example that the presence of the factor of 3 does not change the *proportional* relationship between 'F' and 'a'; but the true value of the x-factor has to be found from further experiments, so that an *exact* relationship can be formulated. In fact, you know from experience that the force must also be proportional to the amount of material in the object (or what we intuitively think of as its weight or 'heaviness'): if you try to push two identical pieces of furniture at once, by yourself, so that you have doubled the amount of matter in the 'object' (that is, you have doubled its 'mass') then you will need to push twice as hard to get the same acceleration as before. Mathematically speaking, this means 'F' is also proportional to 'm', the mass of the object. If 'F' is proportional to both 'm' and 'a', and there are no other factors involved, then you can assume that 'm' is the x-factor in the equation for force: F = m x a. Double either 'm' or 'a', and 'F' also doubles.

That is essentially what Newton's experiments showed: a *quantitative* relationship between the measurable quantities of force (F), mass (m), and acceleration (a), known as Newton's Second Law of Motion. In algebraic equations, usually the multiplication sign is left out, so Newton's Second Law is written more compactly as F = ma. It mathematically *defines* the concept of force.

Incidentally, you might have noticed from experience that

the amount of acceleration of furniture also depends on how slippery the floor is, but this is included in the idea of force: if you push a sofa across polished boards, there is not much friction working against your push, and you do not have to push as hard as you would across carpet. The 'F' in Newton's Second Law is the *effective* force, your push minus the resisting force of friction: $F = F_1 - F_2$, if you want to be technical. Of course, in order to experimentally derive the Second Law of Motion, you would have to do all your experiments on the same kind of floor, so that each time you would have the same effective force. As Galileo showed, when he minimised the influence of air resistance in his tests of the effect of gravity on falling objects, scientific experiments must be designed to take account of all the relevant factors.

In defining the concept of force, the Second Law also defines the concept of weight, which is the common name given to the special force of gravity that pulls us downwards and makes us feel heavy: your weight = your mass x acceleration, the acceleration being the rate at which gravity is pulling you downwards. It is the same rate at which you would fall to the ground if your chair was pulled from under you, because the Earth's gravity is acting on you all the time, pulling you downwards whether you are sitting or standing, running or falling. As Galileo showed, this downwards acceleration is constant – it is the same for all of us, 9.8 metres per second per second when we are near ground level – so our weight is proportional to our mass (in our daily activities on Earth, it is 9.8 times our mass).

Because of this proportionality, the two terms 'weight' and 'mass' are often used interchangeably, if imprecisely: when we say we *weigh* a certain number of kilograms or pounds, we really mean that this is our *mass*, which is *proportional* to our weight. It is not simply the proportionality that causes the confusion; before Newton pointed out the *conceptual* difference between mass and weight, even great philosophers like Descartes had failed to distinguish between them.

The difference between mass and weight becomes clear in space. As we now know from the experimental evidence of astronauts' behaviour, we would be weightless in space, because there would be no (effective) force of gravity pulling or accelerating us towards the floor of the spaceship. (This is because an orbiting spaceship and all its occupants are in 'free-fall', of which more will be said later.) Each of us, however, would still have the same amount of mass (the same amount of matter in our body) as we had back home. The Second Law explains this beautifully: mathematically speaking, no gravitational acceleration means zero acceleration, $a = 0$, so our weight, 'ma', would be $m \times 0$, which also equals 0 (because 0 times anything is 0). On the other hand, once we are back on Earth where gravity *is* acting on us, our mass determines the amount of force needed to pull us to the ground with an acceleration of 9.8 metres per second per second, because $F = ma = m \times 9.8$. The more mass we have, the more force is needed to pull us down, just as more force is needed to accelerate a more massive piece of furniture. Intuitively, we assume that heavier objects will fall faster than lighter ones, but all objects fall at the same rate; it is just that when they hit the ground, the heavier ones do so with a bigger thud, that is, with more force or weight behind them.

Newton also came up with a Third Law of Motion: every

force, or 'action', produces an equal and opposite 'reaction'. If you thump your fist on the table, the table hits back, causing your fist to stop moving and perhaps even to recoil. It will certainly hurt, making you painfully aware of the table's reaction. Newton gave a gentler example: 'If you press a stone with your finger, the finger is also pressed by the stone.'

Mathematically, this law is written as $F_2 = -F_1$, where F_1 is the force of your fist on the table, and F_2 is the force of the table on your fist. Or F_1 may be the force with which you hit the ground after a fall, and F_2 is the force of the ground on you. The equals sign says that these two forces have the same strength: if you fall to the ground with a force of magnitude m x 9.8, this is the same sized force you will feel from the ground as it reacts on you. The minus sign says these two forces act in opposite directions: by convention, F_2 is assumed to be positive, in the upwards direction, while $-F_1$ is in the negative or downwards direction. (In this book, usually I will leave out any minus signs, concentrating on the numerical value of a force rather than on its direction.)

The Third Law explains why earlier philosophers were wrong in assuming that stationary objects have no force acting on them. All objects on Earth have the force of gravity acting on them, no matter what other forces are acting: gravity is acting on a billiard ball rolling across a horizontal table, on an object in free-fall, and on a teapot sitting on a table. Gravity pulls the teapot downwards, but the table pushes up against this force, holding the teapot, motionless, in balance. Without the table to hold it up, the teapot would fall to the ground, gravity then being the only force acting on it (apart from a negligible air resistance, but imagine it falling on the airless Moon to see the point). Once the teapot hits the ground, it feels a new reactive force from the ground itself, which stops it from falling any further, and which might even be strong enough to shatter the teapot.

As for a billiard ball rolling across a table, having been initially accelerated into motion by the force of a billiard cue, it is now rolling along without the need for a continued horizontal force to keep it moving. (This is Newton's First Law.) However, it *is* being affected by the vertical force of gravity; but since this is balanced by the vertical (upwards) reaction of the table, there is no net vertical force acting, and so these forces do not affect the horizontal motion of the ball. Friction and air resistance, not gravity, will eventually slow down the billiard ball. The importance of Newton's laws of motion is that they use precise language that gets to the essence of things.

Newton's importance

Newton is immortal not just because he turned his experimental discoveries about motion into mathematical laws. After all, Galileo and others had laid firm foundations for these laws. Rather, it is Newton's work in theoretical physics that makes him so important. Einstein is the quintessential theoretical physicist for most of us, but he was indebted to his heroes, Newton and Maxwell, both of whose portraits were on his study wall. Maxwell has never received the popular acclaim he deserves, but Newton is still an intellectual icon because of the breathtaking imagination and skill with which he actually *created* both the modern discipline of theoretical physics and much of its mathematical language (which we call 'calculus').

Maxwell would take Newton's language and turn it into high art, creating another scientific revolution in the process.

Back in 1850, however, when the teenaged Maxwell was preparing to go to Cambridge, Newton was beyond compare. His theory of gravity was still the only satisfactory theory in physics, and both physicists and laypeople were in awe of him.

WHY NEWTON HELD THE WORLD IN THRALL

When Newton applied his experimentally based laws of ordinary motion to falling objects, he became the first modern theoretical physicist. I have already spoken about the 'force' of gravity – it is part of common speech these days – but I have been speaking with hindsight. Gravity is nothing like an ordinary push-or-pull force. Unlike the tangible blow from a billiard cue, you cannot see or hear gravity; it is so immaterial, and yet so pervasive, that it took scientists a long time to get a handle on it, to find a suitable name for the way it works. It was Newton who first clearly defined this mysterious phenomenon as a concrete force.

The definition follows intuitively from his First and Third laws. If a teapot is sitting, quite still, on a table, then the First Law says it will only start moving if you apply a force to it. If you could move the table away without pushing the teapot in any way, the teapot would start moving and would fall to the ground, under the influence of gravity. So gravity must be a force. Alternatively, consider the Third Law: You can *feel* a chair pushing against you when you are sitting in it, so you know it is exerting a force on you. From the Third Law, this force must be equal and opposite to the force of your weight on the chair. In other words, your weight – or the gravitational pull on you – must indeed be a force.

But it was Newton's Second Law that clinched it. By

assuming that gravity *is* a force – that it can be described in terms of the Second Law, F = ma – he was able to come up with a remarkably accurate description of what actually happens to objects moving under the influence of gravity. (To a ball thrown obliquely into the air, say.)

Newton went even further than this. He hypothesised that gravity is a universal force, not just an earthly one. He worked out a formula describing how gravity would theoretically behave everywhere in the universe if it actually did extend beyond the earthly environment – a formula that turned out to be accurate to *within a ten-thousandth of a per cent*, and which has enabled physicists to work out the trajectory not just of a ball, but of a spaceship carrying astronauts to the Moon, for example, and of the planets' gravitational orbits around the Sun. But in Newton's day, while some scientists had begun to think of each planet having its own specific type of gravity, the idea that the Sun and other 'heavenly bodies' had the same gravity as the Earth was considered almost sacrilegeous. Most of Newton's contemporaries believed that the 'heavens' – the home of the Moon, Sun and stars – must be governed by very different laws from the Earth, so they did not expect that earthly gravity would exist in the celestial realms. (They believed this partly because of the prevailing Christian cosmology in their culture, in which the heavens had spiritual as well as physical significance, with hell below us and heaven above. A place people aspired to go, after they were dead, so it must, by its very nature, be entirely different from Earth.)

At that time, in the middle of the seventeenth century, the established world view in Europe was one that had been handed down from the ancient Greeks; it pictured all the planets, stars and the Sun itself as though they were attached to rotating 'heavenly spheres' centred on the Earth. It was a harmonious cosmology, according well, in principle at least, with both the self-evident passing parade of heavenly bodies through the sky and a literal interpretation of biblical references to the Sun's motion about the Earth. As far as astronomers were concerned, however, there was a problem: viewed from the usual earthly perspective, the planets' paths look very complicated, so if the imagined 'heavenly spheres' had any physical reality, then they must rotate with the help of an incredibly complex set of invisible gears.

A century earlier, the Polish astronomer, Nicolaus Copernicus, had given a mathematical argument which suggested things would look simpler if the Sun, rather than the Earth, were at the centre of the cosmos; he developed his idea after reading that the ancient Greek astronomer, Aristarchus, had suggested nearly 2000 years earlier that the Sun was at the centre of the universe. Aristarchus's reasons have not survived, and his heliocentric view of the cosmos had not found much favour among ancient astronomers. In particular, they pointed out that if the Earth *were* moving around the Sun, then the relative positions of the nearest fixed stars should change during the Earth's yearly journey, a change which did not appear to occur. (Such a change is called 'parallax'; you can see an example of it just by moving your head and watching how the relative positions of trees, fence posts, furniture or other fixed objects change as your viewpoint varies.) Copernicus was mindful that his own contemporaries would also object to a heliocentric cosmology, on religious grounds alone, but he died in 1543, just after he allowed his ideas to become public in his book, *On the Revolutions of the Heavenly Bodies*.

With Copernicus himself out of the way, and his mathe-

matical calculations being extremely complicated, the re-
vived heliocentric hypothesis did not, at first, make any
significant dent in the prevailing cosmology. Besides, you
just had to look up into the sky and you could see that the
Sun was moving, while the Earth was 'obviously' motionless
(otherwise surely everyone would have been flung off the
Earth as it whirled through space?) Because of the intractable
complexity of the planetary motions as viewed from Earth,
however, Copernicus's view gradually gained influential supporters who spread his ideas, and nearly 75 years after their
first publication, Pope Paul V decided it was time for the
Church to take another look at them. In 1616, he convened
a theological panel that unanimously agreed heliocentricism
was 'formally heretical'. Copernicus's book was withdrawn,
pending 'corrections' to it.

One of Copernicus's most famous champions at this time
was Galileo, and he immediately worried he himself might be
charged with heresy. Sixteen years earlier, his countryman, the
pantheistic mystic and former Dominican friar, Giordano
Bruno, had been burned at the stake for his Copernican and
other heretical views. Galileo had already caused a stir when
he published *The Starry Messenger* in 1610; it was an account
of new observations he had made, using his own telescope –
a recent Dutch invention Galileo had personally improved
upon. In particular, his discovery that Jupiter has moons,
which orbit it rather than the Earth, upset the Earth-centred
view of the universe. But the book – whose first edition sold
out in a few days – was passed by the censors of the Roman
Inquisition. After all, it contained no evidence that the Earth
itself moved. Now, however, Italian scientists were unnerved
by the anti-Copernican edict of 1616.

In 1623, a new pope, Urban VIII, took office; he was

unusually interested in the emerging scientific view of the world, and had long been an admirer of Galileo. He encouraged him to keep on with his work as long as he did not defend Copernicus's hypothesis as a proven fact. But in 1632, Urban felt personally betrayed when Galileo published his pro-Copernican book, *Dialogue Concerning the Two Chief World Systems*, (that is, the Earth-centred system favoured by the Church, and which had been handed down from the ancient Greek astronomer Ptolemy, and Copernicus's Sun-centred system). Brought before the Inquisition, Galileo argued that the *Dialogue* was a debate – a dialogue – but it was evident he favoured the heliocentric side of the argument and he was charged with heresy in 1634.

Approaching 70, and after a lifetime of service to science, he was absolved of the usual 'censures and penalties' for the crime by agreeing formally, and humiliatingly, to deny that the Sun was the centre of the world, to recite the 'seven penitential psalms' once a week for the next three years, and to refrain from all teaching and other public scientific activities. He was helped in this last matter by being sentenced to imprisonment at the Inquisition's pleasure. Urban eventually allowed him to return home under house arrest (where he wrote his most important book, summarising his work on motion and falling bodies. It was smuggled out of Italy and published in Holland as *Two New Sciences*). Meanwhile, the *Dialogue* was banned; it remained officially prohibited by the Catholic Church until 1822. Galileo was formally absolved of heresy in 1992.

However, Galileo's contemporary and admirer, the brilliant German astronomer, Johannes Kepler, had already made a stunning observation, which provided the mathematically minded with powerful indirect evidence that the planets – including the Earth – do move around the Sun. Copernicus's Sun-centred cosmology had been based as much on philosophical grounds as observational ones. Central to this philosophy was the ancient Pythagorean belief that circles, being completely symmetrical and therefore 'perfect', were the only possible shapes for 'heavenly' motion, which must also, by definition, be perfect. So he imagined the planets moving in perfect circles around the Sun. Unfortunately, the planets do not trace out circles, not even around the Sun, so Copernicus's model provided only an approximation of the planetary orbits. But Kepler was a true physicist, another of the founders of modern experimental physics. He left philosophy out of it, and looked only at what was actually happening in the sky; by painstakingly analysing decades of detailed astronomical data about each planet's position in the sky at various times of the year, he discovered that if you looked at it from the Sun's point of view, all the planets did trace beautifully simple paths. They were not circles, though, but ellipses.

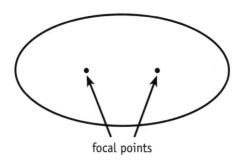

focal points

Unlike circles, ellipses do not just have a single centre, they also have two focal points (or foci). These are not part of the ellipse itself, but are a part of the geometrical scaffolding needed mathematically to construct the ellipse; in the diagram, I have taken away the rest of the underlying construction and left the focal points because they are important in Kepler's scheme. You can gain a sense of how they differentiate an ellipse from a circle by imagining that the curve in the diagram is flexible, so you can gently squash its sides inwards, thus forcing the top and bottom outwards so the whole curve looks like a circle; as you do this, imagine the two foci sliding towards the centre as the sides move inwards, until the foci coincide at the centre and the ellipse becomes a circle:

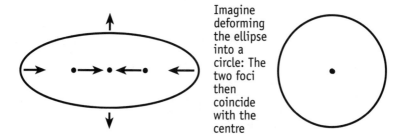

Imagine deforming the ellipse into a circle: The two foci then coincide with the centre

Kepler's celestial geometry was not exactly heliocentric or Sun-*centred*, because he envisaged the Sun being placed not at the centre of a circle but at one of the foci of an ellipse, each planet tracing out an elliptical rather than a circular orbit around the Sun. (The other focal point has no relevance to this astronomical picture, physically speaking.)

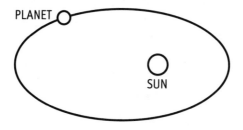

The Moon's orbit around the Earth is also elliptical, but only just – it is a slightly flattened circle:

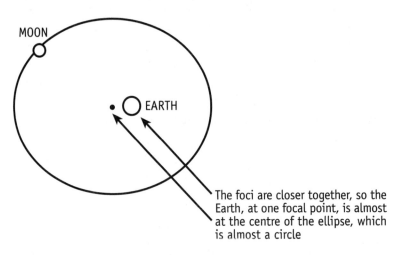

The foci are closer together, so the Earth, at one focal point, is almost at the centre of the ellipse, which is almost a circle

Elliptical curves can be described precisely in terms of relatively simple mathematical equations – unlike Copernicus's still-complicated nest of circles, gears and epicycles – and once you can easily name or describe something, it takes on a sense of reality; heretical or not, Kepler's simple elliptical planetary orbits took on a life of their own in the minds of astronomers and physicists.

Direct physical evidence that the Earth does actually move would not be obtained for another century, when the aberration of starlight (and later, the fabled stellar parallax) was detected. But in Newton's hands, the mathematical evidence alone became irresistible.

How the universal theory of gravity was built: Imagining the world with the language of mathematics

Kepler had managed to *describe* the planetary orbits mathematically, but Newton's goal was to build up a picture of *why* the planets moved in the first place. Like Copernicus, he began by examining the scientific legacy of the ancient past. In the absence of telescopes or space probes, information about the heavens had been built up over the millennia through a mix of unaided observation and speculation. From such a mix had emerged the ancient science of astrology, in which the 'planets' – including the Moon and the Sun, which were thought to be planets in those geocentric days – were assumed to influence human behaviour. This made sense in that the planets' movements through the sky were periodic, as were the cycles in human lives; for example, women know that their moods have a monthly cycle, according to their menstrual period, and the Moon, too, has a monthly cycle, taking 27.5 days to orbit the Earth, and 29.5 days to go from its full phase, then waning to new moon, and waxing to the next full moon. The idea that we might be affected by the 'planets' was therefore a legitimate theoretical hypothesis, of a similar kind to those made later by what we now call theoretical physicists. (Astrologically, the Moon was, indeed, considered to influence the emotions, while the posited effects of Jupiter and Saturn have been handed down to us in the words 'jovial' and 'saturnine'.)

In the hands of reputable practitioners, speaking philosophically if not theologically, the pagan art of astrology was a mathematical and scientific discipline – the precursor to modern astronomy. It required accurate records of the planets' positions in the sky at any time, and horoscopes required detailed geometrical calculations of these positions relative to each other, at any given time. (Apparently, the young Newton was moved to teach himself mathematics in order to read an astrology book he bought at a fair in Cambridge.)

Astrologers even speculated on a physical mechanism that would explain their subject, and they supposed some sort of strange power – we would now call it a force – must be emanating from the planets in order to influence us here on Earth. The Elizabethan mathematician, magus and natural philosopher, John Dee, assumed that all entities in the universe exuded such 'rays', which he thought might be something like those emanating from the dark rock lodestone – rays the ancients had called magnetism. (Dee had been arrested by the Catholic authorities of Queen Mary's time for 'calculating', 'conjuring' and 'witchcraft', accusations based on his work with mathematics and astrology, both of which were widely seen as dangerous arts. He was later released, however, and on Elizabeth's accession to the throne in 1558, he was officially asked to use his astrological skills to choose a propitious date for her coronation. It seems he got it right!)

Dee's younger contemporary, William Gilbert, went so far as to suggest that magnetism held the planets in their orbits, and several decades later, Kepler wondered whether some sort of 'astrological' emanation from the Sun might be the power by which it kept the planets on their elliptical paths. Newton – who was quite at home with the ideas of alchemy, astrology and other 'occult' ways of enquiring into the meaning of the world – developed these ideas by supposing that some sort of invisible *force* was holding the planets in their orbits. His experimental and mathematical work had made the idea of force a concrete and definable one, giving him the language with which to take his predecessors' vague, non-quantitative ideas and turn them into a full-blown theory.

According to his young disciple, William Stukeley, the

chain of events that led Newton to wonder if the posited force from the Sun was that of *gravity* began with the fall of an apple. For some unrecorded reason, this everyday event made him wonder whether the Earth's gravity could be felt not only by the apple but also by the Moon. If so, he mused, perhaps the Moon's circular motion around the Earth was somehow balanced by the pull of the Earth's gravity, which kept the Moon from spinning off into space.

He did not, at first, think of gravity as *causing* the Moon's motion; that idea occurred to him some years later, when his contemporary and rival, the English polymath Robert Hooke, suggested that although the Moon appears to move sideways *around* the Earth, it is also being continually drawn *towards* the centre of the Earth, just as a stone tied to a string and whirled about your hand does not fly off on a tangent, but is continually drawn back to your hand by the force of the tension in the string. After hearing about Hooke's idea, Newton realised that a falling apple, too, is drawn towards the centre of the Earth, and he concluded that the Moon's circular acceleration might have the same cause as the acceleration of the apple as it falls to the ground, namely, the Earth's gravity. Then he took another leap of imagination, suggesting that the planets' motion around the Sun was caused by the *Sun's* gravity.

Nowadays (post-Einstein), we express this idea by saying the Moon is actually *falling* around the Earth, under the influence of the Earth's gravity, while the planets are falling around the Sun. Similarly, a spaceship in orbit is falling, and we say that, like the Moon and planets, it is in 'free-fall', just as an apple falls freely from its tree. In the case of astronauts inside an orbiting spaceship, both they and the spaceship are falling at the same rate, as Galileo could have told us. Con-

sequently, there is no *relative* acceleration between the astronauts and their spaceship, so by Newton's Second Law, there
is no relative or *effective* force pulling them to the floor of
the spaceship. This means that when they jump off the floor
into the air, they do not fall back down again as they would
on Earth, because as they fall, so does the floor. Relative to
the spaceship, they are weightless, and stay floating in the
air until they do a somersault, or push themselves off a wall
or ceiling, or otherwise *force* a change in their motion.

Einstein clarified the relationship between gravity, free-fall
and acceleration in his general theory of relativity, but it was
Newton who first suggested there *was* such a universal relationship. He expressed the idea that gravity is the single
cause of all these different types of motion – the Moon's, the
planets', the apple's – by saying: 'There must be a drawing
power in matter ... like that we here call gravity, which
extends itself thro' the universe'.

However, he did not assume that only large objects like
the Earth and the Sun have gravity. Dee had already supposed that all objects exude magnetic-like 'rays', but Newton
ascribed such a *general* 'drawing power' to gravity, not magnetism. Hooke and others had wondered if the Sun and the
planets had their own specific types of gravity, but Newton
made the formal hypothesis that there is a single type of
gravity, and that it is a property of matter itself, so that all
objects have it, not just the Earth. You and I have it, an apple
has it, the Moon and stars have it, and so we all attract each
other, just as the Earth attracts us. Attraction is the nature
of gravity. In fact, Newton suggested, you and I attract the
Earth itself, in equal and opposite measure to its force on
us: it pulls us downwards, towards itself, while we pull it up
towards us, in accordance with the Third Law of action and

reaction. This seems a ridiculous conjecture on the face of it. We fall straight to the ground, under the influence of the Earth's gravity; we do not see the Earth lunging up to meet us.

But Newton was not speaking in terms of everyday intuition; rather, he was using the language of mathematics. Not only the language of his Third Law, but also that of his Second: in general, this law is written as $F = ma$, where 'F', 'm' and 'a' represent a generic force, mass and acceleration, respectively. In specific applications of this formula, these letters take on specific contexts and numerical values, and if more than one application is being considered, it may be necessary to write the formula using different letters to distinguish between the situations. I want now to use the Second Law to compare the *force of the Earth's gravity on one of us* with the *force of our gravity on the Earth*; I will use the subscripts '1' and '2' to indicate the two different applications of the formula.

The force of the Earth's gravity on one of us is $F_1 = m_1a_1$, where m_1 is our mass, and a_1 is the rate at which we fall towards the Earth (9.8 metres/sec/sec). The motion is downwards, in the so-called negative direction, so if it is necessary to indicate the direction of motion, the 'a_1' term can be written with a negative value, as -9.8 metres/sec/sec, making the force itself negative, too. The force of *our* gravity on the Earth, according to Newton's bizarre-sounding theory, is $F_2 = m_2a_2$, where m_2 is the Earth's mass, and $(+)a_2$ is the amount of its acceleration upwards (in the 'positive' direction) towards us.

The plus and minus signs can be used to indicate the opposite directions of these forces, but now the Third Law comes into it: this law says that, if you ignore the direction and just look at the strength of each of these forces, then they are equal: $F_1 = F_2$, which means $m_1 a_1 = m_2 a_2$. But the mass of the Earth, m_2, is huge compared with our mass m_1, so the only way these two quantities can be the same size is if the huge value of m_2 is counterbalanced by a tiny value of a_2, while the tiny m_1 is balanced by a relatively large value for a_1. It is a similar argument to saying $2 \times 100 = 50 \times 4$, both expressions having the same numerical value, 200: if 2×100 represents $m_1 \times a_1$, and 50×4 is $m_2 \times a_2$, you can see that the large m_2 (50 compared with 2, the value of the small m_1) is counterbalanced by a small a_2 (4, in contrast to the large value of 100 for a_1). In other words, the value of m_2 is 25 times that of m_1, but the value of a_2 is only $1/25$ of a_1.

Of course, the Earth's mass is more than 25 times a person's; it is 10^{23} times a person's mass. This notation follows from the fact that the number '10' is a '1' with 1 zero, and is technically written as 10^1; 100 has 2 zeroes – it is formed by multiplying 10 by 10 and is written as 10^2; 1000 ($10 \times 10 \times 10$) has 3 zeroes and is written as 10^3, while one million is a '1' with 6 zeroes (1,000,000) and is written in shorthand as 10^6. But 10^{23} is a '1' with 23 zeroes after it. To counter-balance such a huge m_2, a_2 is so small as to be negligible, as the following example shows; (the calculation underlying the example follows from $m_1 a_1 = m_2 a_2$ if Newton's special language of calculus is used to define the *concept* of acceleration, so that two additional equations are obtained: 'acceleration $= a_1$' and 'acceleration $= a_2$'). Suppose all the world's six billion people got together on a platform a metre off the ground, which broke under their collective weight so they all fell one metre to the ground under the influence of the Earth's gravity. In that time, the Earth would move towards them – under the influence of *their* gravity –

a distance of less than a trillionth of a millimetre! In our everyday experience, therefore, there is no perceptible acceleration or lunge by the Earth in our direction. Even the mutual gravitational effect we have on each other – pulling us (ever so slightly) towards each other – is imperceptible compared with the relentless, downward pull of the Earth.

While the gravity of everyday Earthly 'objects' like ourselves does not noticeably affect the Earth, Newton realised that the Moon, on the other hand, is large enough to produce potent gravitational effects. He correctly suggested that it pulls the Earth's seas towards it, for example, causing the waters on the side of the Earth facing the Moon at any time to 'lunge' or bulge towards it, producing a high tide; low tides result in other places from where the water is sucked away into the high tide bulge. Similarly, most of the other planets are bigger than the Earth – they are certainly bigger than the Moon – so their gravitational effects on each other are not insignificant, the way an apple's or a person's is: it was the unseen Neptune's gravitational effect on Uranus that was causing the observed distortion in Uranus's orbit.

But while the Moon causes the Earth – most noticeably, the oceans – to 'lunge' or accelerate slightly towards it, the fact that the Moon is so much lighter than the Earth means that the Earth affects the Moon much more powerfully. Which is why the Moon orbits the Earth, because it is actually 'falling' towards it under the influence of the Earth's relatively enormous gravity, just as an apple naturally falls from its tree to the ground. And because the Sun is so much more massive than the planets, its gravity causes them to accelerate or 'fall' towards it, while the planets themselves have only a tiny gravitational effect on the Sun.

The Moon and the planets keep to their orbits rather than

falling right into the Earth or the Sun for the same reason
that an artificial satellite fired from Earth keeps to its orbit:
it is moving too fast to fall straight back down again, because
once it does begin to fall, it has travelled so far that the
Earth's surface has begun to curve away from it; it begins to
fall *around* the curve of the Earth, kept in free-falling orbit
by the Earth's gravity. (Space probes are fired with a greater
speed than satellites so they can break free of the Earth's
gravity altogether and go on to other planets, or beyond the
solar system itself.) You can get a visual idea of this from the
next set of diagrams.

Newton's theory was almost pure speculation: no one had
been to the Moon or launched a spaceship when he made
his revolutionary prediction that gravity is a property of *all*
matter, anywhere in the universe. And that the Earth's gravity is strong enough to extend as far as the Moon, and is
keeping the Moon in its orbit, while the Sun's gravity can
be felt on Earth and is keeping Earth and the other planets
in orbit. Kepler, Dee, Hooke and others had had vaguely
similar ideas, but the reason Newton turned his speculations
into a proper theory was that he was able to precisely *define*
what he meant by his 'emanations' – by his hypothetical,
universal force of gravity. The Second Law defined a general
force, but Newton went looking for a specific definition of
gravitational force that could be applied anywhere in the
universe, not just near the surface of the Earth where $F = m \times 9.8$ applies.

For a start, if he was right in thinking gravity was a
property of all material objects – a property of mass – then
the definition of universal gravity would have to be more
complex than the general definition $F = ma$ because the
acceleration term itself would then depend on the mass of

An orbiting satellite (or moon or planet) is actually 'falling' under the influence of gravity:

If you throw a ball obliquely into the air, its path is curved like this:

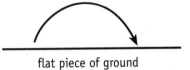

flat piece of ground

The faster you throw the ball, the further it will travel before it hits the ground. If you throw it hard enough, it will go so far that the Earth noticeably begins to curve by the time the ball hits the ground:

The ground looks flat from where you are standing, but someone who is watching your super-human throw from afar can see that the Earth's surface is curving away as the ball falls

If you launch a satellite into the air, it is travelling so fast and far that its curved path looks like this:

The Earth's gravity keeps pulling on the falling satellite so it will continue orbiting the Earth

the attracting body: as I showed earlier, a person of mass m_1 has a relatively large gravitational acceleration (a_1) towards the Earth because of the Earth's huge mass (m_2), while the Earth's acceleration (a_2) towards the person is tiny, in proportion with the tiny mass of the person. This means that the gravitationally induced acceleration of one object towards another is proportional to the second object's mass (and conversely, the acceleration of the second object towards the first is proportional to the first object's mass). I could write this as $a_1 = k \times m_2$, or, more compactly, $a_1 = km_2$, and conversely $a_2 = km_1$, where I am using 'k' again for the x-factor or proportionality factor. (Mathematicians generally choose this letter; it is just a convention, but by consistently using 'k' as the generic symbol for proportionality, mathematicians flag the fact that this kind of proportional relationship exists.)

Newton wanted a general equation for the force of gravity between two objects, and such an equation can be found from the Second Law, by replacing a_1 with km_2 to obtain $F_1 = m_1 a_1 = m_1 km_2$, for the magnitude of the force of gravity acting on the person of mass m_1, and caused by the Earth, which has mass m_2. It is equal and opposite to the force due to the person acting on the Earth, which (replacing a_2 by km_1) is $F_2 = m_2 a_2 = m_2 km_1$. (You can see that it is equal in size because it does not matter which order you multiply numbers — and all these letters stand for numbers, the numerical values of the masses, accelerations and forces involved: $m_2 km_1 = m_1 km_2$, just as $2 \times 3 \times 4 = 4 \times 3 \times 2$.) I have talked specifically about the mutual gravitational interaction between a person and the Earth, but this pair of equations is designed to be a universal law of gravity, so it applies to any two objects in the universe: m_1 might be the mass of the Moon and m_2 the mass of the Earth; or m_1 might be the Earth's mass and m_2 the Sun's; or m_1 might be your mass and m_2 might be mine. Because F_1 and F_2 are equal in size, I can drop the subscripts and say that the general equation for the mutual gravitational force between any given pair of objects of masses m_1 and m_2, respectively, is $F = m_1 km_2$.

Already this formula reveals interesting implications. If m_1

is your mass and m_2 is the Earth's mass, you will fall to the ground with an acceleration $a_1 = km_2$, which we know is 9.8 metres per second per second. Assume for the moment that 'k' is a single, fixed number; if you were to go to the Moon, where you would be subject primarily to the Moon's gravity rather than the Earth's, then the relevant m_2 is now the *Moon's* mass (which is only about one-eightieth of the Earth's mass) and so you would fall to the ground at one-eightieth ($^1/_{80}$) of the earthly acceleration rate, according to the universal equation $a_1 = km_2$.

The gravitational acceleration on the Moon is *not* an eightieth of that on Earth, though, because 'k' is not a fixed number, but depends on another variable factor. According to Newton's final gravitational formula, you would fall about six times slower on the Moon than on the Earth, and this is, in fact, exactly what happens. You have probably seen those famous images of the lunar astronauts leaping into the 'air' and falling slowly, almost gracefully, back down to the ground. On Earth, they would have plummeted very quickly and ungracefully – if they could jump against the Earth's pull in the first place, laden as they were with huge backpacks of equipment. Newton had never seen any such phenomenon, but it was all there in his mathematics.

The variable factor in Newton's final formula is the effect of distance on the force of gravity; it answers the question, What happens to gravity the further away from its source you go? In particular, how strong is the Earth's gravitational effect on the Moon? Imagine trying to answer such questions 350 years ago, when the skies were almost a complete mystery. Newton's approach to the problem was a breathtaking yet relatively simple synthesis of experimental insight and mathematical logic.

Using the well-known observation that the Moon takes 27.5 days to orbit the Earth, and an extraordinary estimate of the distance between the Moon and the Earth made by ancient Greek astronomer-mathematicians, Newton set out to calculate the circular acceleration of the Moon as it orbits the Earth. He used the distance from the Earth to the Moon as the radius 'r' of the Moon's (almost) circular orbit, and the ancient formula 'Circumference = 2πr' then gave the distance around the circle, that is, the distance the Moon travels each time it orbits the Earth. (π – pronounced pi – is a strange number which cannot be written down exactly because it has an infinite number of decimal places. It is approximately equal to 3.14, so the approximate distance around a circle – that is, its circumference – is 2π (or 6.28) times 'r', the circle's radius. You can work out an approximate value of π by measuring with a tape measure the circumference of a known circle – the rim of a teacup or bowl, or a circle drawn with a compass, say – and dividing by twice its radius, that is, by its diameter. No matter what circle you choose, the ratio between its circumference and its diameter is the same; it is always π.)

Once Newton had calculated the approximate length of the Moon's orbit, he could convert the 27.5 days to hours, and divide the distance travelled in a single orbit by the number of hours taken to make the orbit, thereby getting an estimate of the speed of the Moon. He needed the speed to derive the Moon's acceleration, from a formula he had worked out for circular motion.

He then compared the Moon's acceleration with the acceleration due to gravity on the surface of the Earth, concluding (to use modern terminology) that if the Moon *were* actually falling around the Earth, then it was falling much more slowly than things fall when they are close to the Earth's surface. In other words, the force of the Earth's gravity must have diminished enormously by the time it reached the Moon.

Newton also found the same result from a formula for the force on the Moon (or any orbiting body) which he had derived from a mathematical investigation of Kepler's analysis of planetary motion; the formula told him that the force required to keep an object in an elliptical or circular orbit gets weaker the further apart are the object and its 'parent'. The agreement between his Keplerian formula and his Earth/Moon comparison convinced him that the force of the Earth's gravity really is the (only) cause of the Moon's motion. By extension – because Kepler's results were based on the actual planetary orbits, and Newton had hypothesised that gravity was a universal property of all matter – the Sun's gravity must be the cause of the planets' motions.

This meant that in his new equation for gravity, $F = m_1 k\, m_2$, the proportionality factor 'k' would have to be replaced by a new factor that included the weakening effect of distance (call it 'r') in an *inverse* proportion: the *greater* the distance away from the Earth (or Sun or other source of mass), the *smaller* the force of its gravity. The 'r' term would come in as a division rather than a multiplication – something like $F = m_1\, m_2/r$. If you double 'r' (that is, you replace it by 2 x r), you are dividing the formula for 'F' by 2, so its value is halved. In other words, the new force (call it F_2) is $m_1\, m_2/(2 \times r)$, or, to make it easier to see as a fraction,

$\frac{m_1 m_2}{(2 \times r)}$; it can be written as $\frac{1}{2} \times (\frac{m_1 m_2}{r})$, or $\frac{1}{2}$ F, where 'F' was the original force before you doubled 'r'. (To see this, recall the arithmetic of fractions. Start with a fraction, say $\frac{6}{3}$ (= 2), and then get a new fraction by multiplying the denominator by 2: $\frac{6}{2 \times 3} = \frac{1 \times 6}{2 \times 3} = \frac{1}{2} \times \frac{6}{3}$ (= $\frac{6}{6}$ = 1), so the original fraction $\frac{6}{3}$ (= 2) has been halved by doubling the original denominator, 3.) Similarly, if you treble 'r', you are dividing F by 3, and its value is only a third of what it was initially, and so on.

Newton had calculated that this weakening relationship between gravitational force and distance is not just an inverse law but an 'inverse-square' one: if you increase 'r' by a factor, then 'F' is decreased according to the 'square' of that factor. A 'square' is the result of a number multiplied by itself: 4 is the square of 2, because 2 x 2 = 4; 9 is the square of 3, because 3 x 3 = 9, and so on, so Newton's final formula for the universal force of gravity was F = $m_1 m_2 G/r^2$, where r^2 (pronounced 'r-squared') means r x r, and 'G' is a scaling factor needed to make the final relationship between F, m_1, m_2 and r exact. (Such an additional factor was not needed in the Second Law because there is an exact relationship between F, m and a: F = ma.) In other words, the 'k' factor used in the interim formula F = $m_1 k m_2$ (= $m_1 m_2 k$, because multiplication of numbers can be done in any order), turned out to be G/r^2.

The scaling factor 'G' is very small, 0.00000000006673 in metric units; it was measured in laboratory experiments that determined how much gravitational force was produced between two metal balls of known masses, separated by a known distance. As you would expect from the fact that we do not notice our gravitational effect on each other, this force was extremely small, and 'G' is the scaling factor that produces a tiny 'F' from everyday masses like those of metal balls, apples or people. The gravitational force only becomes strong when something as massive as the Sun, the Moon or the Earth is involved (via the m_2 term, which for the Earth is 10^{23} times the 'everyday' mass of a person).

But it is the inverse-square part, the $1/r^2$ factor, that is the distinguishing feature of Newton's equation of gravity. It will have controversial ramifications in Maxwell's story because it turns up as a factor in the laws of electricity and magnetism, too.

It also explains why lunar astronauts fall to the ground at $^1/_6$ of the rate they would fall on Earth, rather than $^1/_{80}$ (in proportion with the Moon/Earth mass ratio). Newton proved mathematically that a spherical object's gravity is concentrated at its centre, just as magnetism is concentrated at the poles of a bar magnet. This means that when standing on the Moon's surface, you are much closer to its source of gravity than you are to the Earth's source when you are standing on *its* surface, because the Moon's radius is much smaller than the Earth's. These radii each represent the 'r' term in the inverse-square law (as applied in each of these particular examples), so if the Earth and Moon had the same amount of mass, then the Moon's gravity measured on its surface would be *larger* than the Earth's (as measured on *its* surface). The force of the Moon's gravity on a Moon-walking astronaut is therefore much larger than expected, considering the Moon has only $^1/_{80}$ as much mass as the Earth.

Newton's theory of gravity in a nutshell

Newton had taken a theoretical idea – that a universal force of gravity causes planets, moons and other satellites to move in their orbits – and had deduced from calculations based on the actual motion of the Moon, and on Kepler's actual planetary ellipses, the mathematical nature of such a force. With the help of the Second Law, he could express his theory mathematically: the planets (and all satellites) are held in their orbits by the force of gravity acting between satellite and 'parent', a force whose strength is given by $F = m_1 m_2 G /r^2$. (In fact, this force acts between any two objects, orbiting or otherwise.) This formula also mathematically expresses Galileo's discovery that heavy objects do not fall faster than light ones: any object has a gravitational acceleration that is independent of its *own* mass m_1, namely $a_1 = m_2 k$ or $m_2 G/r^2$.

All Newton's wild and radical speculations, and all his careful calculations, are expressed in this single equation, which, thanks to the methods of calculus he invented, can be used (when directions are taken into account) to accurately describe all falling motion almost anywhere in the universe, from the path of a ball thrown into the air to the elliptical orbit of a planet or a spacecraft. This is what mathematicians mean when we say mathematics is an elegant, economical language. It is surely a beautiful fulfilment of the ancient quest to encapsulate true wisdom in a single image: to see the universe in a 'grain of sand', or to express the cosmic order in a single word like 'Om'. But the equation has a power the other images do not. It is objectively testable: its descriptions can be tested against what actually happens.

Newton's equation provided far more than an accurate

description of natural events. After all, Kepler had already shown that the actual planetary orbits could be *described* mathematically. Newton's grand achievement had been to take Kepler's ellipses and incorporate them into a mathematical theory about *why* the planets move at all.

The difference between a theory and a hunch

A successful theory is logically consistent, philosophically satisfying and sufficiently physically accurate over a wide range of situations to provide new insights about the physical world. It does not have to be completely accurate, and it does not have to provide the last word on reality, as long as it satisfies these criteria. Newton's theory of gravity provides both an accurate description of the observed facts of planetary motion, and a logically consistent, philosophically satisfying explanation of them, based on the common experience of an everyday force rather than some mysterious 'emanation' or a special celestial type of gravity. It also explains a whole range of other phenomena, like the tides, and the fact that things fall more slowly on the Moon than on Earth, and – as Newton's friend Edmond Halley showed – the motion of the comets. Halley used Newton's theory to deduce that a series of historically observed comets was the same comet, so that (as Newton had suggested) comets are not necessarily one-off phenomena – popularly supposed to be supernatural portents from God – but are astronomical bodies with gravitationally defined paths. He correctly predicted the next few appearances of this comet, which we now know as Halley's comet. Similarly, Maxwell used Newton's theory in making his prediction that Saturn's rings are not solid, and Adams and Leverrier used it to explain the observed distortion of Uranus's orbit, thereby predicting the existence of Neptune. (Astronomers still use the same principle to discover new planets way beyond the solar system.) These and many other predictions and explanations are based on logical deductions from Newton's solid, mathematically expressed *theory*. A hunch, on the other hand, is intuitive rather than logical. It may come up with a right result – and sometimes even for the right reason – but until it is developed into a detailed theory it cannot produce a whole chain of logically derived consequences.

For instance, while Newton was working privately on his theory, the brilliant, cantankerous Robert Hooke also came up with the idea that the planets' elliptical motion about the Sun is due to the Sun's gravity, which, he said, is governed by an inverse-square law. It was a remarkable insight, but Hooke could neither express it in a physically satisfactory way nor prove it mathematically. Like some of his colleagues, he had begun thinking of the celestial bodies as having their own gravity, and although he took a further step by supposing the planets had the same type of gravity as the Sun, he did not imagine gravity as an inherent, *universal* property of matter itself; he thought that only bodies of the same 'nature' could attract each other, and this was too vague and too limited an idea to provide any new understanding of gravity and its role in celestial motion. (For instance, he supposed that comets had a different type of gravity, which caused them to be repelled from the Sun and planets rather than attracted by them; Halley later added support to Newton's simple universal law of attraction for all types of objects, including comets.)

As for the inverse-square law, while Newton derived it rigorously, as a logical consequence of Kepler's observations and his own comparison of orbital and gravitational acceleration, Hooke essentially guessed it. He imagined that the force holding the planets and the Moon in their orbits was something like the light from a candle, and noted that Kepler had suggested an inverse-square law for the intensity of light; the further from the candle you go, the weaker is the intensity of its light. He also 'derived' this law mathematically, but although he got the right answer (probably because he was looking for it), his mathematical and physical argument was incorrect, because he did not properly understand the mathematical nature of force. An incorrect derivation of a correct law does not offer a satisfying explanation of the law, because it does not rightly show the physical and logical chain of cause and effect.

Halley and his colleague, Christopher Wren, came closer to correctly deducing the law when they mathematically calculated that objects moving in a *circular* orbit must be subject to a central, inverse-square force law; although they were unaware of Newton's work at the time, they used the same argument that he initially used when adapting one of Kepler's observations to the (approximately) circular motion of the Moon. Because ellipses are more complicated than circles, it took all of Newton's renowned mathematical prowess to derive the inverse-square law for the *elliptical* paths of the planets, and not surprisingly, Hooke, Halley and Wren were not able to do it – despite an inducement from Wren of a prize of a 40-shilling book, not to mention, he reminded them, the likelihood of fame. They did not yet know that Newton had already scooped them, but knowing his mathematical reputation, Halley looked him up during a visit to Cambridge, and asked him what he thought of the inverse-square idea. When Newton told him of his own calculations, Halley was so impressed he urged him to publish his theory.

Newton was reluctant, partly because at that stage, he still felt his theory needed some mathematical fine-tuning. But his reluctance was also due to the fact that he was still bruised by criticism of his earlier theory of light, by Hooke in particular. Newton had proposed that light travelled from its source to our eyes as a series of corpuscles or particles – rather than as waves, as both Hooke and his Dutch contemporary Huygens supposed – and naturally enough, both Huygens and Hooke criticised the particle hypothesis. But Newton's most important theory about light was that the colours revealed when light passes through a prism are fundamental components of light, rather than the effect of impurities in the glass, or of various mixtures of 'light' and 'darkness', as had been generally supposed. Hooke, Huygens and others vehemently disagreed with Newton's (correct) explanation of colour, and Hooke also claimed that Newton had stolen some of his ideas. It was a claim Hooke would also make when, thanks to Halley's financial backing, Newton finally published his theory of gravity.

In both cases, it seems Hooke was overstating his achievements. He did have an incredible sense of intuition, with which he'd anticipated many theories, including those of light and gravity; his ideas stimulated others, particularly Newton, and in this sense, he made a vital contribution to theoretical physics that deserves the acknowledgement he craved (and which Newton, angered by Hooke's overblown claims and accusations of plagiarism, later tried to deny him). But despite the fecundity of his imagination, Hooke did not formulate any complete, consistent theory. He was a much better experimentalist than theorist, and he made important original contributions in many experimental areas, particularly mechanics, microscopy and biology.

As for Newton's theory of gravity, it was first published in 1687 – twenty years after he had first begun working on the problem of planetary motion – in a great book called *Philosophiae Naturalis Principia Mathematica* (*Mathematical Principles of Natural Philosophy*). Nowadays referred to as 'Newton's *Principia*', it is the most important book in the history of science because it introduced the world to the first consistent, broad-ranging scientific theory.

Newton and his legacy

Newton's simple equation of universal gravity reimagined humanity's place in the universe. We may no longer be able to see ourselves at the spatial centre of the cosmos, but we had become more powerful than we had ever dreamed, because we no longer had to take cosmology on faith. As Maxwell would say, our reason not our faith was now made the judge of God's works.

It was revolutionary stuff because Newton had developed his theory during a time of superstition and religious dogmatism, in which the Earth was officially held to be the centre of the cosmos and comets were widely seen as evil portents. In 1666, the year in which Newton did some of his greatest work (formulating the bases of both calculus and the theory of gravity), the level of scientific sophistication in England was so low that powerful members of the clergy blamed the great materialist philosopher, Thomas Hobbes, for the Plague and the Great Fire of London – signs, they said, that God was not impressed with Hobbes's ideas. (Hobbes wrote that war and the fight for self-preservation are natural human states, so that without laws to enforce peace, our base natural instincts would ensure that 'the life of man [is] solitary, poor, nasty, brutish and short'. He made enemies by arguing in favour of secular rather than religious law for the regulation of these natural instincts, and in 1666 he was banned from publishing in England.)

Newton himself came from fairly humble origins. His father, a successful but illiterate farmer, had died before Newton was born, killed in the English Civil War that continued throughout Newton's childhood. When he was three years old, his mother remarried, but her new husband did not want young Isaac tagging along, and the toddler was left in the care of his grandmother. Some say his traumatic childhood was to blame, dominated as it was by war and abandonment, but whatever the cause, Newton had a rather testy, suspicious nature, which in later life occasionally bordered on the delusional. He was a match for the egocentric Hooke, and it has been suggested that his famous quote about standing on the shoulders of giants may have been a cruel jibe at Hooke as well as a humble acknowledgment of the truth: he made the comment in a letter to Hooke, who, apart from constantly accusing Newton of plagiarism, was short and stooping in stature (not one of the giants whom Newton wished to acknowledge?) I do not know whether Newton really was being malicious in this case, but it would not have been out of character.

However, he could also be generous with his time, money and even his praise. And he was awe-inspiringly brilliant: although he managed to put himself through Cambridge University (by doing a Cambridge Fellow's chores for a few weeks a year, such as emptying his chamber pot, cleaning his room and setting the fire in his fireplace), teaching methods were so haphazard in those politically turbulent times that the great inventor of calculus and the mathematical theory of gravity was virtually self-taught in mathematics.

In 1688, in the aftermath of the Civil War and the Restoration, both sides of Parliament – Whig and Tory, Protestant and Catholic – united in orchestrating a bloodless coup against the unpopular Catholic king, James II, by his Protestant daughter, Mary Stewart, and her cousin and husband, William of Orange. William and Mary had agreed to accept Parliament's Declaration of Rights limiting the power of the monarchy, and on their accession to the British throne in 1689, the world's first modern constitutional monarchy was born. Newton himself sat in that famous 1689 Parliament, having been elected to represent Cambridge University: in addition to having recently published the *Principia* to great acclaim, he had also been vocal in defending the university's rights, such as the independent choice of staff, against the increasingly autocratic James II, who had wanted to staff the universities and other institutions primarily with Catholics.

After this 'Glorious Revolution', the religious and political culture in Britain was more liberal, and more open to a scientific approach to cosmological issues; by 1705, times had changed so dramatically since the dark days of war, superstition and plague in which Newton had grown up, that he was knighted – the first scientist to be so highly honoured.

When he died in 1727, he was buried in Westminster Abbey. Voltaire, the French writer and philosopher, described with awe the pomp and ceremony of his state funeral: 'I have seen a professor of mathematics, simply because he was great in his vocation, buried like a king who had been good to his subjects.' Even in death, Newton cast a huge and glorious shadow, and it took another 150 years for someone else to move into the light beside him.

RITES OF PASSAGE

Maxwell had just turned nineteen as he prepared to leave Scotland for Cambridge. He was a good-looking young man of middle height, with fine features, raven black hair and deep brown eyes. He was noted for having an easy spring in his step, but he still spoke in a distinctive, hesitant way, despite his efforts to follow his old school master's advice: in a letter to his father six years earlier Maxwell had written, 'P– says that a person * of education never puts in * hums and haws; he goes * on with his * sentence without senseless interjections. N.B. Every * means a dead pause.'

Just before he left for England, he attended a meeting in Edinburgh of the British Association for the Advancement of Science, colloquially known as the British Ass. He would later write satirical poems about British Asses, but on that summer night in 1850, he was still shy and inexperienced at such gatherings. Nevertheless, he made an impression on the audience when he addressed them at question time. As later recalled by Professor William Swan, who was present at the meeting, initially there was awkward surprise among the members as they gazed at this 'raw-looking young man who, in broken accents, was addressing them'. But, continued Swan, the young man's embarrassment did not get the better of him, and he 'manfully stuck to his text' until he eventually gained the respectful attention of his hearers.

So much so that after the meeting Maxwell was approached by a tall, confident, handsome fellow with blond hair and brilliantly blue eyes (of which he had always been proud); at 26, he had already gained a reputation as a brilliant mathematical physicist. He was William Thomson, the child prodigy who was on his way to becoming the most famous physicist of his day. (Thomson would eventually become Lord Kelvin.)

Thomson had graduated from Glasgow University at sixteen – the age when most students were just beginning their higher studies – and then took a degree at Cambridge. At the age of 22, he returned to Scotland and, following in his father's footsteps, became a professor at Glasgow, a position he would keep for 53 years. Maxwell's first attempt at public speaking, that night at the British Association meeting, was well rewarded because Thomson wanted to know more about his ideas, and would become one of his closest colleagues.

But first Maxwell had to go to Cambridge – whereupon the odd, shy youth became a popular young man: after an initial settling-in period, he suddenly made 'a troop' of friends, much to Lewis Campbell's 'needless pangs of boyish jealousy'. The Cambridge crowd considered him to be 'genial and amusing', with wide-ranging interests he loved to discuss. A friend recorded in his diary, 'Maxwell as usual showing himself acquainted with every subject on which the conversation turned. I never met a man like him. I do believe there is no subject on which he cannot talk and talk well too, displaying always the most curious and out of the way information.' Some found his quick changes from one subject to another hard to follow, but his personal charisma was such that one of his fellow students remembered taking a walk with him and not understanding a word he said, and yet, recalled the student, 'I would not have missed it for anything'.

Everyone agreed Maxwell had a great sense of humour; he was always writing wryly comic or satirical poems, like 'A Vision of a Wrangler, of a University, of Pedantry, and of Philosophy', which begins,

> Deep St. Mary's bell had sounded,
>   And the twelve notes gently rounded
> Endlesss chimneys that surrounded
>   My abode in Trinity.
> (Letter G, Old Court, South Attics),
> I shut up my mathematics,
>   That confounded hydrostatics –
> Sink it in the deepest sea!
>
> In the grate the flickering embers
>   Served to show how dull November's
> Fogs had stamped my torpid members,
>   Like a plucked and skinny goose.
> And as I prepared for bed, I
>   Asked myself with voice unsteady,
> If of all the stuff I read, I
>   Ever made the slightest use.

Maxwell also liked making clever puns, which 'no one enjoyed more than himself', according to one fellow student by the name of Lawson (who then went on to recall several such puns). But unlike his hearty schoolfriend Tait, Maxwell rarely expressed his amusement with outright laughter; a twinkle in the eye was more his style. His face reflected sensitivity rather than mirth, and Lawson also remembered that 'Everyone who knew him at Trinity [College, Cambridge] can recall some kindness or some act of his which has left an ineffaceable impression of his goodness on the memory – for "good" Maxwell was, in the best sense of the word'. Another friend said he had 'one of those rich and lavish natures which no prosperity can impoverish'. But Campbell summed it up when he said his friend was 'one of the best men who have lived'.

His eccentricity did not seem to diminish his popularity – on the contrary, good friends like Tait and Campbell loved him for it. At Cambridge, it manifested itself in such habits as his passion for exercise. At home at Glenlair, he liked to get up early in the morning, and get his exercise – which he deemed crucial so he would not yawn the day away – by taking the dogs for a walk, collecting fruit, or preparing the horse to bring up the water barrel. But in Cambridge he used to work late, and then do his physical exercise, running up and down the stairs in the early hours of the morning. This understandably annoyed the other denizens, who took to throwing things at him as he passed their rooms until he got the message and took a quieter form of exercise. (Like swimming in the nearby Cam River; he loved swimming, which he did for most of the year, except in the very coldest months.)

While Maxwell inspired unanimous admiration for his personal qualities, he received the same response for his intellectual abilities, the most important accolade from his fellow students being his election as a member of the prestigious undergraduate society known as 'The Apostles'. The society consisted of the twelve students its members considered the most promising in the university. (Famous subsequent members of this group included the philosopher–mathematicians Bertrand Russell and Alfred North Whitehead, the philosopher G.E. Moore, and the writer and publisher Leonard Woolf.)

Discussions with his fellow Apostles helped Maxwell hone his views on the philosophy of language and reality. While physics has replaced philosophy as the primary source of information about physical reality, philosophy still has an important role to play in analysing the way we perceive and describe reality. So much so that physics was called 'natural philosophy' until the end of the nineteenth century. In the seventeenth and eighteenth centuries, the very success of physics, and of pure mathematics itself, had revitalised traditional philosophy, spawning two new branches: rationalism – arising from a faith in mathematical proof – which held that reason is the ultimate guide to the nature of reality; and empiricism – a direct response to the success of Newton's experiment-based physics – which held that sensory experience is the basis of all knowledge about reality.

As a student at both Edinburgh and Cambridge universities, Maxwell studied both these branches: the rationalism of Descartes, Spinoza, Leibniz and Hobbes, and the empiricism of Newton's friend Locke, of Berkeley – perceptive critic of Newton, mathematically speaking, and also of Halley, whom he called an 'infidel mathematician' – and of fellow Scot Hume. He also studied a synthesis of the two from Kant, who believed that even such apparently empirical concepts as space and time are ideas constructed by our minds. In this Kantian vein, Maxwell's lecturers tended to favour the view that science is merely descriptive – it is not the ultimate truth about reality, but works merely by creating useful analogies with which we can approximate, or think about, our perception of physical reality.

Maxwell also read the works of philosophers like George Boole and John Stuart Mill, who wrote on the relationship between language, logic and thought. These philosophical issues interested him much more than the technological applications of physics, although these were very impressive, particularly in the growing electrical communications industry. The first international telegraph cable had recently been laid, under the sea between Dover and Calais, and there was enormous optimism about the possibility of uniting the whole world in this way. There was great potential for electrical theory to be further developed so that it could be applied in this communications revolution, and Maxwell's new friend Thomson was throwing himself into the task. His success would soon make him rich and famous. However, Maxwell's preference for philosophy over technology would ultimately set him above Thomson in terms of both intellectual achievement and lasting fame. With the Apostles, he began to develop his own philosophical ideas.

The Apostles set essay topics for themselves, and they would then discuss their essays. There was plenty of scope for debate because, for the philosophically inclined as well as the technologically minded, the early 1850s were an exciting time to be a student (assuming you were a man, of course. Women were generally barred from attending universities, for reasons we now see as ridiculous: women's potential role as wives and mothers meant it was 'pointless' for them to study; they were biologically incapable of higher learning; it was not ladylike; it was not natural; and so on.) One of the key debates at the time was the relationship between science and religion. (Women's rights would have to wait another two decades to become topical again, with the publication in 1869 of J.S. Mill's controversial *The Subjection of Women.*)

Many scientists, still riding on Newton's wave of success, had turned to science rather than religion for answers to the important questions about the nature of the universe and our role in it. The great French mathematical physicist, Pierre Simon Laplace, had set the tone at the dawn of the nineteenth century: when asked by Napoleon where God fitted into his work, Laplace said, 'Sire, I found no need for that particular hypothesis'. Other physicists, including Maxwell, saw science as a means of understanding, and therefore glorifying, God's creation. (Under the influence of his growing scientific sophistication, however, Maxwell's religion would become increasingly personal and private – a matter of faith, not science – and he later refused to publicly comment on the relationship between the two.)

Science was not the only threat to organised religion at that time; it was also being challenged in a more social context. Marx and Engels had published their *Communist Manifesto* in 1848, and perhaps Maxwell had heard about it at the University of Edinburgh. At any rate, although he was a landlord, and accepted the privileges of his class, he also had a social conscience. At Cambridge, he was inspired by J.F.D. Maurice, a former Apostle who founded the Christian Socialist movement, and who in 1853 was stripped of his professorship of theology at King's College, London, because of his 'heretical' views on the idea of eternal hellfire. (He saw it as metaphorical rather than literal.) Maurice believed in workers' cooperatives and education as the way to maintain a fair society, and would found the Working Man's College in 1854, and later the Queen's College for Women.

In this social and scientific context, Charles Darwin was privately developing his radical theory of human evolution by natural selection, at whose hand biblical literalism would take another battering. He would publish *The Origin of*

Species by Means of Natural Selection in 1859, and his theory would bring renewed heat to debates about God's role in designing and creating the universe. Meanwhile, in an Apostles essay on the nature of the evidence for design by an intelligent Creator, Maxwell wrote (following Kant) that our understanding of nature is limited by the structure of our brains, so that the very belief in design 'is a necessary consequence of the Laws of Thought acting on the phenomena of perception'.

In other words, the existence of mathematically expressed laws of nature did not provide evidence that the world had been *designed* according to these laws; rather, they were evidence of the inseparability of mind and matter. 'The only laws of matter are those which our minds must fabricate, and the only laws of mind are fabricated for it by matter.' (Maxwell did believe in an intelligent Creator God, but he never resorted to sloppy scientific arguments to bolster his belief.) His later work would take this interconnection between mind and matter beyond the Kantian distinction between perception and reality – beyond even the Boolean application of mathematical logic to the 'laws of thought' – by showing the extraordinary ability of mathematical language to go beyond both ordinary rational thought and physical sensibility itself, in its prediction of new physical facts.

A hint of how he did this is apparent in a quirkily original Apostles essay on the use of physical analogies in physical theories: he compared the use of physical analogies in science with the making of verbal puns. Then, as now, punning was a favourite game among students, and Maxwell noted the light-hearted Cambridge tradition of fining punsters and repeaters of puns. But, he continued, the theory of puns is simply the reciprocal of the eminently respectable theory of analogies. His argument is worth elaborating because it not only illustrates his wry humour, but also gives an insight into his original way of thinking about language.

A pun is a play on words, a double meaning; an analogy, on the other hand, is a comparison – for example, electricity is said to 'flow' through a metal wire analogously to the way water flows through a pipe. (The usefulness to physicists of a physical analogy like this is that it offers a way of naming and thinking about something mysterious and invisible, like the movement of electricity, in terms of something concrete and tangible, like flowing water.)

The power of the pun arises because the one word is used to express (or hint at) the two meanings, but the power of the analogy lies in the two specified applications of the one word or concept ('flow' in this case). Maxwell summed up these linguistic specifics by saying that with a pun, 'two truths lie hidden under one expression', so that each 'truth' or meaning has *half* an expression, while in an analogy, each truth or concept must be accompanied by *two* expressions (like water and electricity). Mathematically speaking, the reciprocal of a fraction x/y is y/x, so the reciprocal of a half ($^{1}/_{2}$) is two (2/1). Similarly, the reciprocal of 2 is $^{1}/_{2}$. So, Maxwell said, a pun, which has half an expression per word, is the reciprocal of an analogy, with its two expressions per word. Therefore, he concluded, since the use of puns is considered so 'heinous', we can reason by analogy, and then 'deduce by reciprocation the theory of puns'!

Maxwell's essay was a piece of irreverent nonsense, but it was prescient in its examination of physical analogy in terms of the structure of language: in his later work, he would replace the use of physical analogies – which are the result of prior, preconceived ideas – by the use of pure mathematical language whose unexpected consequences for physical reality come from a place *beyond* conscious thought.

Maxwell's time with the Apostles was an enjoyable rite of passage in a journey that would eventually lead him to rewrite the philosophical rules for doing theoretical physics. The process of graduating from Cambridge was not so pleasurable.

Becoming a wrangler

The academic process at Cambridge revolved around preparation for the final, written examination, called the Tripos, after the three-legged stool on which students had traditionally sat when being examined orally. As the examination time drew near, Maxwell had to put his more wide-ranging intellectual enquiries on hold, in order to study for the mathematics-based exam. (Mathematics was seen as a useful mental preparation for all the professions – not only science but also law, administration and, still the most popular of all, the Church. It was also seen as the most objective way of ranking students because questions could be set in which answers were either right or wrong.)

Those who scored the highest marks in the Tripos were assured of good jobs; they were called 'wranglers', the top student being called the 'senior wrangler'. Tait had been senior wrangler two years earlier. He had telegraphed home with the news, and had written to his former teachers in Edinburgh: 'I'm all in a flutter, I scarcely can utter, etc, as the song has it: I AM SENIOR WRANGLER!' He had also won the Smith's Prize, which was awarded for best performance in a more complex mathematics examination, the Tripos favouring speed and accuracy over problem solving. His and Maxwell's old school, the Edinburgh Academy, had taken great pride in this achievement, and had held a celebration in his honour. Cambridge immediately appointed him a Fellow.

Preparation for the Tripos was arduous; Tait recorded his study hours – usually six or seven hours a day, six days a week – and had noted at one point that he had taken a 'brief respite from torment'. Now it was Maxwell's turn, and he expressed his frustration in a 96-line rhyming attack on the exam system, called 'Lines written under the conviction that it is not wise to read Mathematics in November after one's fire is out'. It begins,

> In the sad November time,
> When the leaf has left the lime,
> And the Cam, with sludge and slime,
> Plasters his ugly channel,
> While, with sober step and slow,
> Round about the marshes low,
> Stiffening students stumping go
> Shivering through their flannel.
>
> Then to me in doleful mood
> Rises up a question rude,
> Asking what sufficient good
> Comes of this mode of living?
> Moping on from day to day,
> Grinding up what will not 'pay',
> Till the jaded brain gives way
> Under its own misgiving.
>
> Why should wretched Man employ
> Years which Nature meant for joy,
> Striving vainly to destroy
> Freedom of thought and feeling?
> Still the injured powers remain
> Endless stores of hopeless pain,

When at last the vanquished brain
Languishes past all healing.

[Then he speaks of dreaming of a 'tempting spirit' selling
fame and glory and 'flimsy' (academic) robes rather than
wisdom. The academic temptress, a 'learned-looking maid',
says:]

'Those that fix their eager eyes
Ever on the nearest prize
Well may venture to despise
Loftier aspirations.
Pedantry is in demand!
Buy it up at second-hand,
Seek no more to understand
Profitless speculations.'

Thus the gaudy gowns were sold,
Cast off sloughs of pedants old;
Proudly marched the students bold
Through the domain of error,
Till their trappings, false though fair,
Mouldered off and left them bare,
Clustering close in blank despair,
Nakedness, cold, and terror.

Maxwell was at a considerable disadvantage in the Tripos,
not only because he preferred original thinking to swotting,
but also because of his penchant for making careless mathematical mistakes. Despite his professors' efforts to help him
overcome this flaw, he never quite managed it – his mind
was too quicksilver. Even at the height of his career, his
famous German colleague, Gustav Kirchhoff, would say, 'He
is a genius, but one has to check his calculations before one
can accept them'.

Nevertheless, in the winter of 1854, Maxwell emerged from the Tripos as 'second wrangler'; he tied for the Smith's Prize with the senior wrangler, E.J. Routh. (Routh went on to become a mathematician, and a successful teacher, training 27 senior wranglers.) Maxwell's father wrote to him in his characteristically brief but touching way: 'Tonight or on Monday I will expect to hear of the Smith's Prizes. I get congratulations on all hands [for the Tripos result] ...' Two days later, he had the news:

> [Your cousin George] came into my room at 2 am yesterday morning, having seen the Saturday *Times*, received by express train, and I got your letter before breakfast yesterday. As you are equal to the Senior [Wrangler] in the champion trial, you are but a very little behind him.

Now freed from academic study, Maxwell was ready to let his imagination soar, and to become a true 'natural philosopher'. But first he had to learn the ropes, and William Thomson would become an indispensible guide.

A FLEDGLING PHYSICIST

Maxwell's notorious mathematical carelessness was one of the reasons that, despite his examination success, he was not immediately appointed a Cambridge Fellow. (The other was that he needed to 'pay attention to the classics'.) Even in later life, Maxwell openly admitted his carelessness: 'I am quite capable of writing a fancy [incorrect] formula and not finding it out till I come to work at it.' He knew, however, that although precision and accuracy are vital in the end, real mathematical skill is linguistic and imaginative rather than arithmetical: mathematics is a language for thinking dramatic new thoughts, not merely for doing accurate book-keeping.

Maxwell spent his time after graduation either at home at Glenlair, or as a private tutor at Cambridge. One of the first things he did after leaving university was write to Thomson, in February 1854. Over the previous decade, Thomson had published some important papers on electricity, and it was on that subject that Maxwell saw him as a mentor. He asked what reading he should do to get himself properly up to date on the topic, which had intrigued him since his boyhood trips with his father to meetings of the Royal Society of Edinburgh. In approaching Thomson like this, Maxwell showed an instinctive understanding of the little-known fact that, for the most part, academic scientists share their knowledge very generously.

To facilitate this generosity, which in turn facilitates the development of new ideas, active competition between researchers is discouraged. An unwritten protocol says that if someone has declared an intention to develop a particular idea, it is not acceptable for others to take up that idea – to 'steal' it. (On the other hand, it is perfectly acceptable, indeed, complimentary, to take someone's ideas and develop them further in another direction, providing due acknowledgment is given to the original source of the ideas.) So researchers are free to dash off ideas and questions to each other, and their published papers often contain not only a citation of relevant published works by other authors, but also acknowledgments of conversations with colleagues who have offered important original insights into the research at hand.

Luckily for us, Maxwell and his colleagues had no choice but snail mail for their communications with each other, and it has left its luminous trail so that, 150 years later, we can get a glimpse into history in the making. A glimpse that shows that scientific stereotypes do not reflect the lives of most scientists, even great ones. Maxwell was not a 'lone genius', and his letters show us not only how intellectually indebted he was to colleagues like Thomson, but also how the distractions of daily life interfered with his work while enriching him as a person. (Newton, on the other hand, *was* pretty much a stereotypical, antisocial, lone genius. He did not like company, he never married, and he would become so engrossed in his work that he would forget to eat, or comb his hair.)

Thomson responded to his young colleague's request for advice, and by November, Maxwell was able to write him a long letter, beginning, 'I have heard very little of you for some time except thro' [Cambridge teachers] Hopkins and

Stokes, but I suppose you are at work in Glasgow as usual. Do you remember a long letter you wrote me about electricity, for which I forget if I thanked you?' He then set out several detailed pages on the state of his understanding of the subject, and ended with a list of questions and a heartfelt plea: 'This is a long screed of electricity, but I find no other man to apply to on the subject so I hope you may not find it difficult to see my drift.' Isolated at Glenlair for much of the time, Maxwell was largely dependent on Thomson's generous replies to get him started in the field he would ultimately make his own.

Not that he was working solely on electricity – he always had several scientific projects on the go, and at this time he was also working on both the theory and practice of optics. He invented a device for looking into the retina of the eye, practising on his dog, and he expressed the wonder of his discovery in another letter to Thomson, saying of the dog's eye, 'This is really a beautiful object and by no means difficult to be seen. The dog does not seem to object.'

Concern for the dog is typical of Maxwell. He never became so involved in science that he forgot about its subjects – or his family and friends. At the height of his career, when he had the clout to get away with it, he would take his dog Toby everywhere, even into Cambridge's Cavendish Laboratory; he would impishly point out that Toby was 'thoroughly conversant with the details of the Laboratory and some of its apparatus'. On a more serious level, the scientific challenge for Maxwell in this context – the challenge to be both objectively open-minded and humane – is powerfully summed up by Lewis Campbell:

> I remember his once speaking to me on the subject of vivisection. He did not condemn its use, supposing the method could

be shown to be fruitful, which at that time he doubted, but – 'Couldn't do it, you know,' he added, with a sensitive wistful look not easy to forget.

Early in the new year, his father, now in his mid-sixties, became seriously ill, and Maxwell took a few weeks off to do 'a little cooking and buttling' for him, as he told his friend Cecil Munro; he said that when his father was 'on his pins again', he would come down to Cambridge, and that he was looking forward to 'working off a streak of mathematics which I begin to yearn after'. In a second letter to Munro, however, he betrayed a hint of frustration with the generation gap:

> I have no time at present for anything except looking through novels, etc., and finding passages which will not offend my father to read to him. He strongly objects to new-fangled books, and knows the old by heart. But he likes the 'Essays in Intervals of Business', cause why, they have not too many words.

His father recovered, and Maxwell was able to get back into science. By September, 1855, a little more than eighteen months after he had left Cambridge, he had organised his electromagnetic knowledge to the point where he wrote another letter to Thomson, laying out his electromagnetic influences. Firstly, Faraday, whose *Experimental Researches in Electricity* Maxwell had read soon after he had graduated. Secondly, Ampère, who had mathematically described Oersted's discovery that electric currents create measurable magnetic effects. (When we speak of electric current in terms of 'amps', we really mean 'ampères', a unit named in honour of Ampère.) Finally, Thomson himself – in particular, the paper he had written when he was just seventeen, in which, inspired by an idea of Faraday's, he showed that certain electrostatic effects could be described mathematically in the same way as had been done (by the Frenchman, Jean-Baptiste Fourier) for the flow of heat as it radiates from a heat source.

Electrostatic effects are caused by the attraction and repulsion of electrically charged objects, the same kind of interaction that happens with magnets. Thomson's paper suggested that the influence of one charged object on another could be visualised as some sort of 'flow' of electricity between the objects. Ampère had used a different analogy, in which he had imagined that magnets and electric charges emanate disembodied *forces*, in the same way that Newton had visualised the Earth emitting gravity; these remote forces actively pushed the electrical objects together (or pulled them apart), while Thomson's 'flow' analogy suggested a more fluid mechanism of interaction. Thomson had hoped that by using a different approach from Ampère, he might be able to throw more light on a subject that still had not yielded to a complete mathematical description, despite Ampère's brilliant work – and also that of Ampère's German disciple, Wilhelm Weber.

Ampère, Weber, Thomson – these were the giants whom Maxwell now wished to join. His goal was ambitious. To do for Faraday's discovery – that a moving magnet somehow starts up an electric current in a nearby loop of wire – what Ampère had done for Oersted's: describe it mathematically. He planned to take the mathematical approach used by Thomson rather than Ampère, because Thomson had taken his lead directly from Faraday's work. Just to make sure it was all right with Thomson, he told him what his plans were, and asked Thomson if he had already done such research. He was doing the right thing, saying, in effect, You can go first if you want to; I acknowledge your priority in developing the basis of these ideas.

But Thomson was becoming sidetracked on his more practical electrical investigations – so much so that he would soon become a director of the Atlantic Telegraph Company – and Maxwell was able to write to his father, 'I got a long letter from Thomson ... and he is very glad that I should poach on his electrical preserves'. In October, he received more good news: he had been made a Fellow of Trinity College, Cambridge, which meant he had a proper job. (It also meant he had to leave his father and base himself at Cambridge, so he still kept an eye out for academic positions closer to home.)

In November, inspired by J.F.D. Maurice, he began teaching evening classes for working men in Cambridge (and he also spent time nursing his sick friend Pomeroy). He taught a class of mechanics – teaching such things as decimal fractions, and the mechanical principles underlying the use of levers and the work done by machines. He found the men eager to learn, and would give up a night a week for many years to teach such classes. He also supported the workers' agitation for shorter working hours and more education, and at home at Glenlair, he and his father initiated a reading program for the estate's workers and their children. He loved literature himself, and sometimes mentioned in his letters what novels he was reading. One letter to his father at this time mentioned Charlotte Brontë's *Shirley*, William Thackeray's *The Newcomes*, and, hot off the press, Charles Kingsley's *Westward Ho!* He was particularly impressed with Charlotte Brontë's work, as he mentioned in a letter to

Campbell; he thought she was a writer who 'continued to think and acquire principles'.

By the end of the year, he was able to settle in at Cambridge and complete his first electrical paper, which represented the first step towards his goal of mathematically expressing Faraday's discovery. But science did not replace life. When Pomeroy died a year later, Maxwell wrote to a mutual friend,

> It is in personal union with my friends that I hope to escape the despair which belongs to the contemplation of the outward aspect of things with human eyes. Either be a machine and see nothing but 'phenomena' or else try to be a man, feeling your life interwoven, as it is, with many others, and strengthened by them whether in life or death.

This letter is all the more poignant because a few months before it was written, his beloved father had died, at the age of 66. Having seen father and son together a couple of years earlier, his fellow student Tayler had noted how touching it had been

> to witness the perfect affection and confidence which subsisted between [them]: the joy and satisfaction and exulting pride which the father evidently felt in his son's success and well-earned [local] fame; and, on the other hand, the tender, thoughtful care and watchfulness which James Maxwell manifested towards his father.

Characteristically, the 24-year-old Maxwell, who also still grieved for his mother, chose a poem (called 'Recollections of Dreamland') as the means to express his pain, at losing not only her but now also his father and best friend – the father who had lovingly nurtured his talents through childhood but who would not live to see them in full bloom:

O my heart is hot within me, for I feel the gentle power
Of the spirits that still love me, waiting for this sacred hour.
   Yes – I know the forms that meet me are but phantoms of
      the brain,
For they walk in mortal bodies, and they have not ceased from
      pain.
   Oh! those signs of human weakness, left behind for ever
      now,
Dearer far to me than glories round a fancied seraph's brow.
   Oh! the old familiar voices! Oh! the patient waiting eyes!
   Let me live with them in dreamland, while the world in
      slumber lies.

ELECTROMAGNETIC
CONTROVERSY

When his father died, in April 1856, Maxwell told Campbell he had a personal mission to carry on his father's work 'of personally supervising everything at home'. As for his own pursuits,

> It was my father's wish, and it is mine, that I should go on with them. We used to settle that what I ought to be engaged in was some occupation of teaching, admitting of long vacations for being at home; and when my father heard of the Aberdeen position he very much approved. I have not heard anything very lately, but I believe my name is not yet put out of question. If I get back to Glenlair I shall have the mark of my father's work on everything I see ...

A few days later, he wrote to Thomson with the news, saying his father had been calmly prepared for his death.

Later in the year, Maxwell was, in fact, given the position of Professor of Natural Philosophy at Marischal College, Aberdeen. He found it a bit stuffy at first, and wrote to Campbell, 'No jokes of any kind are understood here. I have not made one for two months, and if I feel one coming I shall bite my tongue.'

In 1858, Maxwell married Katherine Dewar, the daughter of the principal of Marischal College. She was seven years older than he, a strictly religious woman of 35 when he married her, and she was not a popular choice among his friends, who thought she did not like to see her husband spending too much time yarning with them. But Maxwell was in love; he wrote to his new wife on an occasion when his university work had temporarily separated them,

> Oft in the night, from this lone room
> I long to fly o'er land and sea,
> To pierce the dark, dividing gloom,
> And join myself to thee.
>
> And thou to me wouldst gladly fly,
> I know thee well, my own true wife!
> We feel that when we live not nigh,
> We lose the crown of life.

Throughout their married life, the Maxwells enjoyed many shared pursuits, like reading the classics together, and riding their horses, and Katherine sometimes helped Maxwell with his physics experiments. She nursed him through both smallpox and a potentially fatal streptococcal infection (the latter acquired when he scraped his head on a tree branch while riding a new horse), and he always maintained she saved his life. In turn, he nursed her through frequent bouts of indisposition, often sitting up with her all night.

Two years before his marriage, in 1856, Maxwell's first electrical paper had been published. It was an interesting time, electrically speaking, because in spring 1855 a telegraphic cable had been laid between London and Crimea, a part of the Ukraine that was then the site of a bloody war between Russia, and Turkey, France and Britain. Florence Nightingale had recently become famous for her attempts to publicise and redress the appalling state of army medical care in Crimea, and the British public were reading for themselves about the shocking Allied casualties through eyewitness accounts published in *The Times*, sent via telegraph by W.H. Russell, the first war correspondent in history to use the new technology. Public opinion in Britain turned against the war, which ended in early 1856.

The Times' instant accounts of the horrors of war meant that telegraphy was on everyone's mind, and in the universities a theoretical controversy was gaining momentum – an argument about the fundamental nature of electricity itself. It could be produced in a battery or generator, and applied in the telegraph, but without a better understanding of how it actually got from one place to another, practical, long-distance telegraphic transmission problems could not be solved. It was becoming crucial – politically and economically, as well as scientifically – that physicists sorted out their differences on how electricity was transmitted, even if they did not yet know what it was actually made of.

The two dissenting views on the matter were the 'Newtonian' and the 'Faradayan', the former being the established view, which used gravitational analogies as Ampère had done. Thomson's heat-flow analogy, on the other hand, was 'Faradayan'. Newton's approach had been used successfully in physics for nearly two centuries; Faraday's idea, however, was only two decades old, and before Maxwell's new paper was published, no one but Thomson had given it more than a passing glance.

History of the controversy

While Newton had been knighted less than 20 years after he had published his theory of gravity, this well-deserved recognition did not mean that his theory had had no contemporary critics. Newton himself had been concerned, because, if a gravitational force were holding the planets in their orbits – a force which, he thought, must be something like the force required to keep an object moving on a circular path when it is tied to one end of a string and whirled around by someone holding the other end – then where was the 'string' connecting the planets to the Sun as they whirled around it? Most known forces were due to direct contacts, like a push or pull, or the tension transmitted down a string, so the idea that gravity is a force was physically (although not mathematically) somewhat problematic to Newton – and to many others.

But the incredible success of the theory of gravity soon made the problem a non-issue, particularly to Newton's followers, if not always to Newton himself: gravity was simply a new kind of force, which did not require direct contact. It acted *at a distance*, just as a strong magnet can attract a nail from a distance. Because the effects of these forces seem to occur instantly – the apple begins to fall to the ground immediately it leaves the tree, and the nail begins to move toward the magnet as soon as they are placed near each other – Newton's advocates assumed there is no transmission process involved at all: gravitational and magnetic forces act *instantaneously*; they do not take any time to travel from the source to the 'target' – from the Earth to the apple, from the magnet to the nail.

This is the essence of the 'Newtonian' view of gravity, electricity and magnetism: they are all imagined to be invisible forces between separate objects or *particles* – forces that affect only the particles themselves, and which act *instantaneously*, *at a distance*, without the need of a physical connection (a string or any other intervening matter) to aid their transmission. In fact, Newton supposed that gravity could leap across completely empty space, and he correctly suggested that the space between the Sun and the planets – 'outer space' – is a vacuum, empty of virtually all matter, including air.

However, the idea of empty space was another controversial aspect of Newton's theory in the early days, and despite its practical success in describing planetary motion and other gravitational effects, the theory was slow to take off outside Britain. After all, the Continental Europeans had their own great thinkers. Their most influential philosopher at that time was the founder of modern rationalism, Frenchman René Descartes; he had produced the most comprehensive system of natural philosophy since Aristotle, and had been one of Newton's most important inspirations.

Then there was Newton's contemporary, German Gottfried Leibniz, who independently invented calculus at around the same time as Newton did, and with whom the increasingly suspicious Newton embroiled himself in a dishonourable dispute over priority. The verdict of historians is that Newton invented calculus first, but that although Leibniz had seen or heard something of it during a visit to England where he met Newton's colleagues, Leibniz's calculus was nevertheless independent of Newton's. So much so that it is Leibniz's more transparent symbolism we generally use today, not Newton's. Leibniz also *published* his version first. But while the polymath Leibniz did original scientific work, particularly in the fields of optics and mechanics, he was ultimately a rationalist philosopher and mathematician, rather than a physicist, and he did not apply calculus to physics the way Newton and his followers did.

Leibniz intensely disliked Newton's idea of gravity as an invisible force capable of acting across empty space – after all, he said, why would God allow empty space at all? Surely

He would show His omnipotence by filling up space with things He had created. Leibniz therefore imagined that the space between the planets and the Sun was not empty, but was filled with infinitesimal, universally interconnected monads or atoms of vital, active force.

Similarly, Descartes had imagined that outer space was filled with vortices of heavenly matter, which buffeted about the planets and caused their paths to twist and bend in their journey through the skies; he assumed – anticipating Newton's First Law of Motion (albeit for metaphysical rather than physical reasons) – that without bumping into this swirling matter, the planets would keep moving in 'perfect' straight lines. He rejected Kepler's suggestion of a force emanating from the Sun and holding the planets in orbit from a great distance, considering it too 'magical' to take seriously.

But Descartes, too, was a philosopher rather than a physicist, like Newton, or his contemporary, Galileo. Like Leibniz's monads, Descartes's vortex model did not have any physical evidence to back it up; it was just an imaginative idea, and his belief that God was the cause of motion itself did not add anything to people's understanding of it. Similarly, Leibniz's dismissal of Newton's theory on the grounds that it reduced God's role to that of a watchmaker rather than a miracle worker, because it allowed the planets to move in mathematically predetermined orbits rather than divinely impelled ones, owed more to religious philosophy than a clear-sighted observation of nature itself. Leibniz was a wonderfully optimistic and visionary philosopher, whose concept of harmoniously interacting monads expressed his belief in the basic goodness and harmony of the world, but he did seem to be excessively confident that he understood

God's purpose. Newton, too, was religious, but he believed his own theory illuminated God's work rather than diminished it. Half a century later, in his philosophical novel, *Candide*, Voltaire mercilessly satirised Leibniz's optimistic faith that God had created the 'best possible world'. On the other hand, he was very impressed with Newton.

Through the influence of Voltaire and other Continental supporters – including Voltaire's feisty lover, Emilie de Breteuil, Marquise du Châtelet, who made the first and only French translation of the *Principia* – Newton eventually replaced Descartes and Leibniz in European natural philosophy. The apparent impossibility of an invisible gravitational force acting instantaneously across empty space was soon forgotten. While his European contemporaries had had a spiritual and intellectual horror of the void and had filled it with all manner of complicated hypothetical substances, the very simplicity of Newton's theory eventually became an important part of its appeal for later generations of philosophers and physicists. An appeal that, when multiplied by the extraordinary accuracy of the theory, had enticed the best mathematical minds of the eighteenth and early nineteenth centuries to apply Newton's mathematical definition of force, and his action-at-a-distance approach to gravity, to the emerging sciences of electricity and magnetism.

Then an outsider appeared on the scene: Faraday, the experimental genius who was self-taught, so he knew almost nothing of mathematics and the established methods of theoretical physics. His lack of education meant he was incapable of being dazzled by the mathematical elegance and simplicity of Newtonian theory, and when he discovered electromagnetic induction in 1831, he did not even attempt to describe it in mathematical symbols, but built up his own mental picture of how a magnet interacted with the nearby wire in which it induced an electric current. A picture that fitted with his experimental results so naturally that he became convinced the mainstream explanation was inadequate.

Like the clear-sighted child in the fable, Faraday reminded physicists that the emperor had no clothes, that the very idea of action-at-a-distance seemed to embody a logical contradiction. He took his inspiration from Newton himself, pointing out that the great man had said, 'That gravity should [be such] that one body can act upon another at a distance through a vacuum, without the mediation of anything else ... is to me so great an absurdity, that I believe no man [of] competent faculty of thinking can ever fall into it'.

While Newton had avoided confronting this problem by confining himself to describing correctly the ultimate *effect* of the force of gravity, Faraday began to look more closely at how this and the electromagnetic forces might actually be *transmitted*. He suggested that while there might be no need of matter (such as air or vortices) to facilitate this transmission process, there must be *something* physical going on in the space between a magnet and a wire, say, in order that the force from the magnet reaches the wire so that electromagnetic induction can occur. He called this 'something' a 'field', and he imagined it sent out invisible 'strings', which he called 'lines of force', through the entire space around the magnet. The magnetic force from the magnet did not leap magically *over* the space directly to the wire, but travelled *through* the space, along these lines of force.

It was Faraday's idea of lines of force that had prompted Thomson to compare electrostatic effects with heat flow, in which 'lines of heat' had been identified. In his first electrical paper, Maxwell had taken Thomson's idea and extended it, conceiving electrostatic lines of force to be analogous to the streamlines you can see in a river as the water flows and swirls into eddies and vortices. It was, in a sense, more akin to Descartes's vision of space than Newton's, but Maxwell was well aware that he had taken a side in a scientific controversy.

During his intense period of preparatory reading on the subject of electricity, he had decided to read about Faraday's ideas first, so he could see them for himself through the eyes of the man who had discovered electromagnetic induction. Only when he had digested Faraday did he read Ampère and Weber, who were among the greatest modern exponents of Newtonianism. Maxwell thought that Ampère's work in particular was brilliant — he called him the Newton of electricity — but he wondered, in a letter to Thomson, if Ampère's mathematical expression of Oersted's result were not a bit contrived: it was a fabulous edifice, but somehow Maxwell could not quite believe in it. Could not quite *feel* the connection between Oersted's physical phenomenon and the mathematical, action-at-a-distance description Ampère had created.

In light of Maxwell's childhood fascination with bell-wires (the secret of which was that by pulling a bell-rope in one room, the pull was transmitted to a bell in another room not by some magical, remote action-at-a-distance, but by a physical 'string' or bell-wire), it is not surprising that he was drawn instinctively to Faraday's field idea. Thomson's influence was also vital, and Maxwell would later acknowledge this in his ultimate treatise on electromagnetism, saying it was to Thomson's 'advice and assistance, as well as to his published papers, [that] I owe most of what I have learned on the subject'. It was Thomson's work, he said, that had convinced him that the disagreement between the Newtonians and Faraday 'did not arise from either party being wrong'. After all, the Newtonian methods worked, and therefore must be correct as far as they went, while in the electrostatic case, Thomson had shown that Faraday's field lines were not vague and unscientific as his critics suggested, but could be described mathematically, analogously to the well-accepted lines of heat flow. But in the more complex electromagnetic case (of which telegraphy was an important application), no one had been able adequately to visualise and mathematically express Faraday's field lines. The problem in this case, Maxwell thought, was that neither Faraday nor the Newtonians 'were satisfied with each other's language'.

Faraday had expressed his field idea in ordinary language, a fact which seems to have prejudiced most of the mathematical physicists of the time, who believed its content was as unsophisticated as its presentation. Maxwell was almost alone in recognising that, 'although not exhibited in the conventional form of mathematical symbols', Faraday's description of the electromagnetic 'field' was profoundly mathematical in conception. His ultimate goal was to translate Faraday's words into more precise mathematical language in the hope that the Newtonians would take it more seriously.

It was an ambition that put him on the losing side at that point in the battle, a brave move for a young physicist looking to make his mark in the established physics community: a community that greatly honoured Faraday for his experimental work, but thought he had stepped out of line with his attempt at theorising. The astronomer royal, Sir George Airy (who had been Lucasian Professor of Mathematics at Cambridge, just as Newton had been), summed up the British scientific establishment's opinion when he said, 'I declare that I can hardly imagine anyone ... to hesitate an instant in the choice between the simple and precise [Newtonian] action, on the one hand, and anything so vague as lines of force, on the other hand'.

Maxwell's genius lay in the fact that he did hesitate, rather than rush along with the sophisticated mainstream. He had called his first electrical paper 'On Faraday's Lines of Force', and it created sufficient interest in the physics community that news of it reached Maxwell's old friend Tait in Belfast, where for several years he had been professor of natural philosophy at Queen's College. (He would also meet and marry his wife there.) But while Tait heard about Maxwell's controversial paper through the grapevine, Maxwell did not rely on chance for Faraday to take notice of it, and boldly sent him a copy.

At 65, Faraday, who had been debilitatingly depressed over the scientific establishment's rejection of him as a theoretical physicist, was overcome with emotion. He replied to the 25-year-old Maxwell,

> I received your paper, and thank you very much for it. I do not say I venture to thank you for what you have said about 'Lines of Force', because I know you have done it for the interests of philosophical truth; but you must suppose it is work grateful to me, and gives me much encouragement to think on. I was almost frightened when I saw such mathematical force made to bear upon the subject, and then wondered that the subject

stood it so well. I send by this post another paper to you; I wonder what you will say to it ...

A working-class dreamer

Faraday had always wanted to be a 'philosopher', a thinker; it galled him to be seen merely as an experimenter, because that smacked too much of the tradesman he used to be. As a young man, he had been intensely idealistic about scientific theorists (or natural philosophers), believing them to be dedicated to the truth rather than to their own petty, self-serving interests, and it had been a bitter discovery to find that even they had their prejudices. He said of the controversy between his field idea and the Newtonian view, 'Everyone blindly follows the Newtonian and ignores mine, even though Newton was no Newtonian'. He hoped that if he 'kept his temper and worked hard', he would finally gain attention as a theoretician, but even during one of his most hopeful periods – when Thomson had first shown interest in his lines of force – he had tempered his enthusiasm by adding, 'But we shall see how the maggot bites presently'.

Faraday was the son of a blacksmith; he was born in 1791, the third of four children in a close-knit but impoverished family that did not have much truck with education. They were members of a small religious sect, originally a breakaway group of Anglicans and Presbyterians called Sandemanians (after the sect's founder, Robert Sandeman); they believed in maintaining a childlike faith, based on the self-evident miracle of the created universe and on the teachings of the Bible. Consequently, they were not interested in formal education, but were concerned with simplicity in language, believing most of the Bible to be literally interpretable because it was written in what they called 'plain style'. (The perennially humble Faraday would remain a devoted Sandemanian all his life.)

In the new age of steam engines, it became increasingly hard for a blacksmith to find work, and the Faradays sometimes went hungry. At the age of thirteen, young Michael went to work at a bookbindery owned by a man whose kindness was destined to make history: George Riebau, a refugee from the French Revolution, who encouraged his young apprentice to spend his spare time trying to read the books he was binding. Faraday soon became captivated by the *Encyclopaedia Britannica*'s detailed entry on electricity, which fired his imagination to the extent that every day he hovered longingly in front of an old rag shop, until he could afford the seven pennies needed to buy two old jars that, with the aid of a bullet and a bit of wire, he turned into his first electrical apparatus.

At that time, the opening years of the nineteenth century, the electric battery had just been invented by Volta. (The terms 'voltage' and 'volt' were named in his honour.) The battery was about to revolutionise electrical research because it created a steady, safe (low-voltage) supply of electric current. When Faraday discovered the *Encyclopaedia Britannica*, however, Volta's battery had been around for only a couple of years, and many scientists still relied on the old way of producing electricity – by using friction to produce static electricity, just as schoolchildren do when they rub their rulers on their jumpers and then try to attract shredded pieces of paper to the electrified ruler. Or when you brush your hair in a darkened room and see the sparks of static electricity. (Static electric and magnetic effects are those emanating from *stationary* charges or magnets. It is the field lines associated with these effects that Thomson was the first to describe mathematically. In an electric current, by contrast, the electricity is continually *moving*, flowing like water molecules in a river; similarly, you have to *move* a magnet in order to induce an electric current in a nearby loop of wire. This is the electromagnetic effect Maxwell wanted to describe mathematically.)

Static electricity could be stored, albeit at sometimes dangerously high voltages, by electrifying a 'condenser' (called a Leyden jar by its Dutch inventor, van Muschenbroek.) Producing weird effects with stockpiled static electricity, like igniting alcohol with an electric spark or causing people's hair to stand on end, had been a favourite and occasionally dangerous eighteenth-century party trick.

It was a Leyden jar that Faraday created from the old jars he had bought. With Riebau's encouragement, he began to conduct his own experiments at the back of the shop, after work; he soon found he had such a flair for experimenting that he set out to reinvent himself as a scientist – a seemingly impossible dream for a young tradesman who could barely read. However, his path was set in 1809 when he began binding and reading a book by Isaac Watts called *The Improvement of the Mind.*

Watts recommended several strategies: keeping a notebook, corresponding with friends to practise expressing ideas, and attending lectures or discussion groups. Faraday became a zealous note-taker, and from 1810, when he was nineteen, he attended weekly meetings of the City Philosophical Society at the home of John Tatum, a silversmith and amateur chemist. According to a later official document on Faraday,

His Master [Riebau] allowed him to attend Tatum's Chemical

Lectures delivered in Dorset Street, Salisbury Square; of these he took copious notes, which he transcribed fairly before he went to Bed. The admission Fee to each lecture was a shilling, and he hoarded up all the Money he got given him, to pay it.

His language being that of the most illiterate, induced Mr Magrath (the Secretary of the City Philosophical Society ...) who attended the same lectures, to devote two hours every week to his instruction; and for 7 years, did Faraday uninterruptedly receive them.

Faraday used his craft to bind his account of Tatum's lectures into a beautiful book, which, lying on the counter of Riebau's shop, caught the eye of a customer, William Dance. Dance was a member of the prestigious Royal Institution, whose premises were not far from Riebau's, and he was so impressed with young Faraday's enterprise that he gave him free tickets to hear several lectures at the institution by the famous chemist, Sir Humphry Davy. This was a great gift, because Faraday, having become extremely interested in chemistry through Tatum's lectures, and through reading Jane Marcet's popular book, *Conversations on Chemistry*, had been sorely disappointed that he could not afford the price of admission to Davy's lectures.

It was now 1813, and Davy was considered to be the best scientist in Britain at that time. The French were impressed with him, too, and despite the fact that Britain and France were at war, the Institut de France had just awarded him the Bonaparte Prize. He had also recently received a British knighthood for his scientific work, and would be elected president of the Royal Society of London in 1820 – extraordinary achievements for a man who also had risen from an apprenticeship (with an apothecary–surgeon) rather than a university degree.

Faraday was fascinated by Davy's lectures; he took detailed notes, and then bound them into a book, which he sent to his new hero. He also asked him for a job. His apprenticeship with the kindly Riebau had ended and he had had to take on a position with a less supportive 'master'; he despaired of ever being able to afford to devote his time to science, especially now that he had to help support his mother and younger sister after his father's recent death. But Davy – who, like Dance, was impressed with Faraday's initiative and grasp of scientific ideas – was able to offer him a lowly position at the Royal Institution. (The newly appointed Faraday travelled to France with Davy, as his valet, when Davy went to collect his medal from the Institut de France.)

After his first promotion, from laboratory sweeper to Davy's scientific assistant, Faraday was completely at home at the institution: his new position also offered him lodgings there, with 'as many coals and candles as he wanted'. Faraday's dream had come true, and to cap it all off, the window of his new room looked out onto Jacques Hotel, where sometimes there were grand parties with such 'excellent' music that he told a friend, 'I cannot for the life of me help running, at every new piece they play, to the window to hear them'. (Faraday had a fine bass voice of his own, and he also played the flute, but his musical interests gradually succumbed to his all-consuming interest in science.)

He married another devout Sandemanian, Sarah Barnard, in 1821, and they would live at the institution until 1858, when Queen Victoria offered them the use of one of her houses. But Faraday retained the humility of his Sandemanian beliefs all his life, even to the point of refusing the queen's offer of a knighthood and a burial plot in Westminster Abbey.

Nevertheless, the brilliant and humble Faraday was very personable; he had had to be, in order to overcome the class prejudices of some of his colleagues. His early career had been blighted by false accusations of plagiarism, and even Davy had jealously turned against him. By the time he received a copy of Maxwell's paper, however, he had become one of the world's greatest scientists (although the scientific establishment's rejection of him as a theoretician hurt deeply). He was also one of the most popular scientific speakers in mid-nineteenth-century London – in contrast to young Maxwell, one of the best scientific writers, but a poor speaker because of his hesitant, heavily accented speech, and his tendency to go off on a tangent as his mind leapt ahead of his prepared notes. Faraday could not afford to take such risks; he had worked hard to emulate the style and sound of successful speakers, hiding all traces of his working-class origins by paying a professor of elocution to attend his early lectures, and to point out afterwards any errors of grammar or pronunciation he might have made.

Faraday's appeal as a lecturer was his use of plain, relatively accessible language. However, the man who danced around his laboratory and hugged his assistant when his experiments were going well was anything but plain in the style of his delivery. Eyewitness Cornelia Crosse commented on 'that wonderful mobility of countenance so peculiar to him', while Juliet Pollock spoke of his gleaming eyes, the hair streaming out from his head, his moving hands and irresistible eloquence. 'His audience took fire with him,' she said, 'and every face was flushed.' As one of his colleagues put it, 'his Friday evening discourses were sometimes difficult to follow. But he exercised a magic on his hearers which often sent them away persuaded that they knew all about a subject of which they knew but little.'

Faraday also instituted regular Christmas lectures for children, of whom he was enormously fond. They responded to him, sometimes running around the block after they met him on the street, so they could come back and have him greet them again. It was a deep sadness to him that he didn't have children of his own. (Neither would Maxwell or Thomson.)

Faraday and mathematics

Faraday's famous 'plain style' of lecturing did not include such abstruse language as mathematics, and not one mathematical symbol appears in the notebooks in which he recorded his experimental results. He regretted his lack of mathematical education, saying, 'All my mathematics consist in that rough natural portion of geometry which everybody has more or less'. However, in light of the dogmatism of most mathematical physicists in persisting in their Newtonian views on electromagnetism, it is not surprising that he also said, 'Sometimes I am glad I am not a mathematician'.

But he was extremely impressed with the open-minded Maxwell. While the established theoretical physicists ignored Faraday's attempts at theorising, Maxwell responded enthusiastically; in November, 1857, almost a year after they had started corresponding, Faraday wrote saying how grateful he was for Maxwell's letters because he had never before communicated with 'one of your mode and habit of thinking ... I hang on to your words because they are to me weighty ... and give me great comfort'.

However, he asked if it would be possible for Maxwell to

express his own work in 'common language' as well as in mathematical 'hieroglyphics', so that he could understand it. After all, he said, 'I have always found that you could convey to me a perfectly clear idea of your conclusions, which, though they may give me no full understanding of the steps of your process, give me the results ... so clear in character than I can think and work from them'.

At first, Maxwell did not believe that the great Faraday could not engage with mathematics. His private response to Thomson was, 'What a painful amount of modesty he has when he talks about things which may possibly be of a mathematical cast'. But when he found out that one of the greatest physicists of all time was virtually self-taught, Maxwell realised, unlike most of his colleagues, that what Faraday lacked in formal mathematical training, he more than made up for with freshness of perspective.

Faraday's fields versus Newtonian particles

Faraday's 'lines of force' were a hypothetical geometrical analogy with the well-known patterns made by iron filings when they line up in the presence of a magnet. If you randomly scatter iron filings on a sheet of paper, and then put a bar magnet under the paper, the filings will coalesce into a beautiful pattern of lines, each of which appears to curve out from one end of the magnet and travel down into the other end, as shown in the diagram below.

This pattern occurs because each iron filing acts as a tiny magnet with a north and south pole. The poles of a magnet are at either end of it, and they 'contain' most of the magnetic force of the magnet; generally, the larger the magnet, the stronger the magnetic force emanating from its poles, so each tiny iron filing is too weak to pull the bar magnet

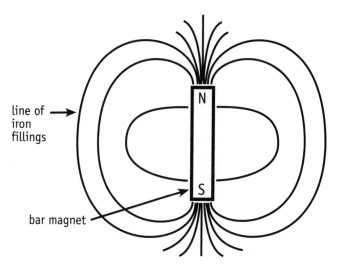

line of →
iron
fillings

bar magnet

noticeably towards it (just as an apple is too weak to noticeably gravitationally attract the Earth) and instead it moves
towards the bar magnet (just as the apple falls to the
ground). Scattered filings close to the ends of the large bar
magnet will bunch up, their opposite poles pulled right up

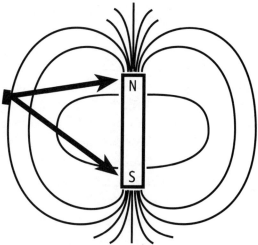

The heavy arrows drawn
from this single iron
filing illustrate the
directions of the
magnetic forces pulling
the south pole of the
filing towards the north
pole of the bar magnet,
and the north pole of
the filing towards the
south pole of the bar
magnet

to the bar magnet by its strong attractive force. Filings that are further away, however, will be attracted in some measure to each pole of the large magnet, and will align themselves so that their north pole faces the bar magnet's south pole, and vice versa.

You also have to take into account the repulsive forces between similar poles, which for simplicity's sake are not shown here; together, all these forces cause each filing to orient itself in a specific way, as you can see in the following diagram.

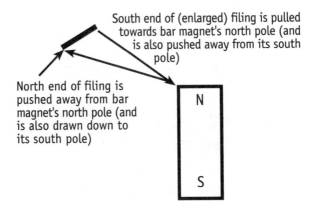

South end of (enlarged) filing is pulled towards bar magnet's north pole (and is also pushed away from its south pole)

North end of filing is pushed away from bar magnet's north pole (and is also drawn down to its south pole)

Groups of filings in the various regions about the magnet also cling to *each other* to form lines, because the north and south poles of adjacent filings are attracted to each other as well as to the large magnet. (They form single lines because only their ends contain the magnetic poles – there is no significant attraction between the *sides* of the filings.)

As far as the Newtonians were concerned, there was nothing more to the lines of filings than an interesting pattern created by the combined effects, frozen in time, of all the remote magnetic forces acting between the bar magnet and the filings. Faraday saw it differently. He was struck by the fact that, under the influence of the bar magnet's strong magnetism, randomly scattered iron filings coalesce into these *particular* curved lines; he pictured the bar magnet's force as being *transmitted* along these lines, from filing to filing, just as a breeze blowing across a farmer's field moves every stalk of wheat in turn.

For the Newtonians, action-at-a-distance forces affected only the particles involved, but it occurred to Faraday that even when the iron filings are not there, the magnetic force from the bar magnet still exists; after all, surely a magnet does not lose its power just because there is no other magnet nearby to demonstrate the effects of that power. He imagined it magnetising or 'polarising' the surrounding space, filling it up with an invisible pattern of curved lines of force similar to those made by the iron filings. The magnetised space was the field, and Faraday imagined the magnet's force was continually transmitted through the field along the invisible lines of force, just as it had been transmitted along the lines of iron filings; then, when any other magnet is placed nearby, it immediately feels the effects of this continual emanation of force. Faraday also imagined that the *number* of lines of force in any given region gives a measure of the force in that region; at the poles of a magnet, for example, there are more lines of force than at the sides.

It was quite an elaborate conception, and it would be a huge task to translate it into conventional mathematical language; by contrast, the action-at-a-distance hypothesis did not require an intricate description of how the magnetic force was transmitted through space, point by point, millimetre by millimetre, because it did not require the force to be transmitted at all: instead of a pre-existing field, there was simply an instant, remote connection between the bar magnet and any nearby magnet. In the case of the iron filings, there was a remote force between the bar magnet and *each* iron filing.

Because each magnet, large or small, has two opposite magnetic poles, each of which affects both poles of other such magnets, the resulting force between two magnets is actually a combination of the separate 'simple' forces acting between the various poles, as shown in the diagrams above. The combination of these simple forces produces an overall force that reorients the magnets as well as pushing them apart or pulling them together, as you can see from the highlighted iron filing in the diagram. But in principle, at least, mathematically calculating the forces on the iron filings is beautifully simple: there is no need to describe an entire, complex pattern of field lines, and to calculate their density; you simply apply – to each pole interaction of any filing – a single equation that describes the remote magnetic force between *any* two magnetic poles. Such mathematical simplicity was a powerful argument in favour of the Newtonian viewpoint, because physicists tend to assume that the simpler of two alternative theories about nature is more likely to be the correct one, other things being equal.

To compound the appeal of the Newtonian approach, the equation describing the 'simple' magnetic force acting between two separate poles is identical in form to Newton's equation for the gravitational force between two objects: $F = m_1 m_2 G / r^2$. (Recall that the objects' masses are m_1 and m_2, respectively, and 'r' is the distance between them.) A hundred years after Newton had proposed his law, the French physicist, Charles Augustin de Coulomb, worked out a way to measure directly the force between two electrically charged objects, and deduced what is now called Coulomb's law of electrostatic force: any two small objects, each charged with static electricity, exert an (equal and opposite) electric force on each other that has the same pattern as Newton's law: $F = q_1 q_2 K / r^2$, where 'K' is a constant (like 'G'), the 'q_1' and 'q_2' terms represent the respective amounts of electric charge on the two objects, and 'r' is the distance between them. Coulomb's law holds for separate magnetic poles, too, with 'q_1' and 'q_2' in the equation replaced by 'p_1' and 'p_2', the magnetic strength of the poles, and an appropriate scaling factor included.

Since Newton's equations worked so well in describing virtually all that was known about gravity at that time, and since Coulomb's law worked for static electric and magnetic phenomena, most mathematical physicists were supremely confident in applying the Newtonian, action-at-a-distance, particle-based model to their descriptions of electromagnetic phenomena – to the electric forces induced by moving magnets, and the magnetic forces induced by electric currents – which also appeared to act instantaneously at a distance.

As Galileo had shown, however, physical intuition is not reliable enough to be the sole guide in the development of a physical theory, and in the absence of definitive physical evidence about whether electromagnetic phenomena really *are* propagated instantaneously, physicists were actually making a *philosophical* choice between strange, invisible 'fields' and remote but familiar forces. Only further physical evidence would reveal which analogy was closest to the actual physical reality of electromagnetism, but the apparent mathematical simplicity of the Newtonian approach had seduced most physicists into accepting it unquestioningly.

Even those like Gauss, Weber and Kirchhoff – who questioned the *physical* interpretation of action-at-a-distance and wondered about the possibility that electric or electromagnetic effects did not propagate instantaneously – used Newtonian *mathematical* methods to express their hypotheses. After all, Newton's theory of gravity was still the only complete mathematical theory of physics, and it made sense to emulate his methods.

It seems Maxwell was the only one who fully understood that different mathematical *language* was necessary if physicists were adequately to express these alternative hypotheses. As for Thomson, who had been field theory's first mathematical champion, he was not as good a philsopher as he was a mathematician.

The favourite and youngest of four children, who had been lovingly and indulgently raised by their academic father after their mother had died when Thomson was six, he was handsome and charming, with 'a singularly winning smile' according to one of his students. He inspired respect and affection in his friends, and he was brilliant. But he could have learned something important from his correspondence with young Maxwell, who carefully set out for him all the papers that had influenced his knowledge of electromagnetism – papers he had studied carefully before making up his mind about his own research direction. Thomson was too independent – too impetuous, too opinionated – to properly absorb other people's ideas, even Faraday's. His younger namesake, Sir Joseph John (J.J.) Thomson (who discovered the electron in 1897), would say he was 'a good radiator and a bad absorber ...' As a consequence, he was an original thinker, but was too quick to jump to scientific conclusions – and too apt to stick to them – to achieve the deep philosophical and mathematical synthesis needed to complete the mathematical description of Faraday's field theory.

Thomson also had a competitive streak. He told Tait – whom he had met after Tait transferred from Belfast to nearby Edinburgh in 1860 (beating Maxwell for the position because he was considered to be the better lecturer) – that he felt his and Maxwell's separate work on electromagnetism was a 'race'. Given the prodigious number of other topics Thomson was working on at the time, perhaps his interest in field theory waned when he sensed Maxwell was 'winning'.

Maxwell, on the other hand, was blessed with a supremely calm disposition; he was not distracted by competitiveness or impetuosity, but calmly investigated all the facts and then went about integrating them into an original philosophical whole. The key to his philosophy, which he had been honing since his Apostles days, was his appreciation of the profound relationship between language and reality. In a contemporary article contrasting Newton's new science with Descartes's vortices, Voltaire had spoken of the 'furious contradictions' between them; the contradictions between Faraday's fields and Newtonian particles, acting-at-a-distance, were no less vehement, but using the language of mathematics in a new and philosophically radical way, Maxwell would eventually resolve them.

MATHEMATICS AS LANGUAGE

Mathematics and ordinary language

When Galileo and Newton used mathematics as the language of physics, they used it as a tool – a servant of physics. In Maxwell's work, however, mathematics would become the 'queen of science', because his physics hinges on the fact that mathematics is a language in its own right. While there are fundamental similarities between mathematical and ordinary language – both are based on definitions and rules of grammar – the uncanny predictive power of physical theories like Maxwell's lies in the *difference* between mathematics and ordinary language.

On the face of it, this difference is one of precision. While ordinary language qualitatively expresses every known concept, mathematics is especially good at expressing quantitative ideas (which is why it is so useful to physicists). Ordinary language also enables you to speak quantitatively to some degree. You can talk about a few things or many things, for instance, or large and small things, and you can precisely specify these quantities and sizes in words – a dozen eggs, 200 dollars, a five-kilometre walk, a three-kilogram weight loss, and so on.

However, we do not normally talk about things like the amount of current flowing when we switch on a light. We do not even speak about the strength of the force of gravity acting on us, even though it is exactly the same thing as our weight. For everyday purposes, it is not necessary to understand anything more about our weight than that it is a measure of how heavy we are. We have no need to define precisely the terms 'heavy' and 'weight', or to wonder why we feel heaviness in the first place – namely, because the Earth's gravity pulls us downwards with a specific, quantifiable force. In ordinary language, force is defined qualitatively – for example, 'power; exerted strength or impetus; intense effort'. The force gravity exerts on us as it pulls us down is a mathematical concept, which is precisely defined by Newton's equation.

You could write this definition in words rather than mathematical symbols, but you would be using those words mathematically, according to mathematical definitions and grammatical patterns, rather than in an everyday way: 'The force of the Earth's gravity on an object equals the mass of the object multiplied by the acceleration produced by gravity, which is proportional to the mass of the Earth and inversely proportional to the square of the distance of the object from the centre of the Earth.' This kind of attention to quantitative, measurable detail, expressed according to the grammatical rules of arithmetic, distinguishes mathematics from ordinary language, and we do not speak like this normally – it sounds stilted and rigid. At this level, though, the difference is one of degree or emphasis, rather than anything more fundamental, and you can certainly see that mathematics is also a language. Like ordinary language, it consists of definitions whose meanings have to be assimilated, and whose grammar provides a structure for ordered thinking and communication. It takes effort to learn these rules and definitions, but even our own native languages need continual work from us if they are to reveal their full richness.

For instance, when reading a literary novel, many of us have to consult dictionaries occasionally for the meaning of some of the words. This can be an imposition, but it can also be a challenge. And it is not just the relatively simple challenge of superficially acquiring a new word. As with mathematical definitions, there are often interesting, sometimes difficult *ideas* that must be assimilated, beyond the specific meaning of the word. Ideas about philosophy or ethics or culture, perhaps – such as with the words 'ontological' and 'solipsistic', for example, or 'anomie' and 'shibboleth' – or about the language itself, such as whether there is a subtle difference in meaning between 'timorous' and 'timid', or whether the difference is primarily one of aural or visual aesthetics. And does some hidden linguistic history underlie the fact that 'dissemble' and 'disassemble', say, are so similar in form but not in meaning?

In mathematics, by contrast, form is extremely important, because the symbolism of the language is an integral part of its content. Consequently, when mathematics is written not in words but in mathematical symbols, the difference between it and ordinary language becomes more obvious, and more fundamental.

For example (to take an illustration from the great literary science writer of the early twentieth century, D'Arcy Wentworth Thompson), thinking in terms of ordinary words, how would you differentiate between the shapes of a rainbow and a hanging chain whose ends are fixed at the same height above the ground? Or between a rainbow and a jet of water from a hose, squirted obliquely into the air so that it forms a beautiful, shimmering arc as all the water droplets fall to the ground? Perhaps you would say they are all curved lines, or arcs, although the rainbow and the water jet always appear upright, while the chain always hangs down, like an inverted rainbow.

In fact, these shapes are all different, although most of us would be hard-pressed to describe or even discern any significant differences using the language of common speech. Mathematically, though, we can be precise: the rainbow is a semicircular arch, the hanging chain forms a unique curve called a catenary, and the shape of the water jet is a parabola; each of these subtly different shapes has a precisely different mathematical definition in the form of an equation. The rainbow's shape is $y = \sqrt{(r^2 - x^2)}$; the water jet is described as $y = -ax^2 + bx + c$; and the hanging chain is $y = a\cosh(x/a)$. As Thompson put it, when you think in terms of mathematical symbols as well as words, 'thought itself is economised', because the symbolism enables you to see at a glance patterns and generalities, similarities and differences, which may not be obvious if you think only in words.

For a more immediate illustration of this point, think of the multiplication tables. All schoolchildren learn these by rote – a good learning technique in this case, because the effort of memorising is repaid by the subsequent effortlessness of working out, for example, the cost of six kilograms of flour when you know that one kilogram costs $1.30. You do not have to ask yourself what multiplication actually *is*, because you just know it works. You do not have to remember the fact that it is a shorthand way of doing addition: 6 x $1.30 is an economical way not only of *writing* $1.30 + $1.30 + $1.30 + $1.30 + $1.30 + $1.30, but also of *thinking* about it.

Sometimes, mathematical symbolism economises thought so much it takes even physicists by surprise – a fact that is at the heart of Maxwell's resolution of the field controversy.

For the moment, $E = mc^2$ is the most accessible example: you can see immediately that energy (E) is in some sense equivalent (=) to matter (m), but without the equation to guide their thinking, few had dreamed of such a bizarre idea. After all, energy – like the power in light and heat – seems almost ephemeral compared with matter – like rock, wood, water, or even our own bodies – and physicists had always believed they were quite separate things. But Einstein had not devised his equation to interpret or describe the nature of matter and energy as observed with human eyes or touched with human hands. He was not looking at physical energy and matter, but was playing around with the theoretical possibilities suggested by his 'special theory of relativity', which he had invented to explain something else entirely: the effect of relative motion on the way people perceive the world, particularly the way they perceive light, which enables them to see the world and make physical measurements of it. (The central hypothesis of the theory is that light travels at the same constant speed relative to everyone. By contrast, a car travels at a certain speed relative to the road, and relative to people standing on the road, but it is not moving at all relative to the people sitting in the car.)

If special relativity theory was right about light, though, then it should also be correct in its prediction that energy and matter are interchangeable, because the prediction was a logical, mathematical consequence of the hypothesis about light. As everyone now knows, $E = mc^2$ *is* correct. Its most famous verification came in 1938, when the Austrian-born physicists, Lise Meitner and Otto Frisch, showed that it explained the process of nuclear fission, which had just been discovered, accidentally, by the German physicist, Otto Hahn. Fission is the process of releasing energy when the nucleus of a radioactive atom like uranium is split, and some of its mass is converted into energy, courtesy of $E = mc^2$. The 'c²' term in the equation is a proportionality or scaling factor, like 'G' in Newton's equation of gravity; 'G' is an extremely small number, but 'c²' is huge, because only a small amount of radioactive matter is needed to release a huge amount of energy. However, mathematicians often prefer to rescale the *units* of energy, so that no scaling factor is needed in the equation itself, which becomes $E = m$. Now the symbolism of the equation clearly shows that energy and mass are equivalent. (A simple example of this kind of rescaling involves changing the units of speed from kilometres/hour to metres/second.)

Einstein's equation also explains nuclear fusion, in which energy is released when atomic nuclei fuse; this is the way the Sun produces its light and heat, when nuclei fuse under its enormous gravitational pressure. And the equation describes the *spontaneous* nuclear reaction which physicists now know causes radioactivity – the phenomenon in which certain material substances (like pitchblende) actually radiate heat and light. Radioactivity had been discovered a few years before Einstein discovered his equation – by the French experimental physicists, A. Henri Becquerel, and Marie and Pierre Curie. (Marie coined the term radioactivity in 1898.) Becquerel, and mathematical theorists Henri Poincaré and F. Hasenöhrl, had each anticipated Einstein's 1905 prediction that energy has mass (so that when a substance like pitchblende gives off energy its mass actually decreases). But in 1907 Einstein suggested the reverse is also true – that mass has energy – so that all mass is *fundamentally equivalent* to 'a store of energy' of amount $E = mc^2$.

The important thing about this is that the equation, the

language, had come first; only later did experimental physicists discover that it described something real. It demonstrates the ability of symbolic mathematical language to go beyond objective description or subjective interpretation of the manifest physical world. In describing something *not* yet manifest, like nuclear fission, in a way that turned out to be an extremely accurate and objective description once the phenomenon did become physical, Einstein's equation seemed to go beyond conscious thought itself. Certainly Einstein himself was not prepared for the consequences. After the nuclear bomb was dropped on Hiroshima, he was devastated. Later, he said, 'If I knew they were going to do this, I would have become a shoemaker instead of a physicist'.

A language of pattern

Crucial to Maxwell's resolution of the field controversy was his intuitive commitment to the idea – which Einstein's $E = mc^2$ later demonstrated so dramatically – that mathematics is not just a descriptive language, in the same way that ordinary words are used to describe things we see or imagine, but that its *linguistic structure* seems to accurately reflect hidden, often unimaginable *physical* structures. I have spoken of mathematics as a quantitative language, but it is much more than that. At its core, it is not so much about numbers themselves as about the relationships between them – the structural and numerical patterns they make.

For a simple example, consider multiplication again. Although the use of calculators may have made our memories of the multiplication tables themselves somewhat rusty, no doubt we can all remember the experience of learning them out loud in class. It was a bit like chanting poetry, because of the rhythmic structure and repetitious patterns in both poetry and multiplication tables. The five and ten times tables were the easiest to learn, because their repetitions are so obvious: the fives end in either five or zero, in an alternating pattern, while the tens all end in zero. Those who enjoy spotting and exploiting these patterns show an ability to think mathematically, especially when they are able to generalise their discovery – to say, for instance, that ten times *any* (whole) number is just that number with an extra zero on the end of it.

In physics, numerical relationships between physical quantities also produce symbolic patterns, as Newton's and Coulomb's laws show. $F = m_1 m_2 G/r^2$ and $F = q_1 q_2 K/r^2$ are visually similar, and they express the same underlying *grammatical* pattern: 'multiply three numbers (m_1 x m_2 x G, or q_1 x q_2 x K), and divide by the square of a fourth number (r)'. This algebraic pattern corresponds to the physical structure of the basic gravitational and electric forces between two objects: namely, that their distant mutual effects are proportional to certain physical attributes of the objects (their masses or charges), and are inversely proportional to the square of the distance between them. But mathematically speaking, the simple grammatical pattern itself is more fundamental than any quantitative physical meaning that may be ascribed to it.

Humans have an apparently innate fascination with patterns, perhaps because our own bodies are essentially symmetrical; or because we needed to learn to recognise the rhythms of the seasons in order to plan the gathering or agricultural activities that provide our food. Whatever the reason, we define ourselves, as a species, largely by our ability both to create abstract languages and to appreciate patterns, so mathematics is the quintessential expression of a defining aspect of what it means to be human.

It is, therefore, a truly international language, whose symbolism reflects its trans-cultural heritage and character. While Latin letters have been used in Newton's, Coulomb's and Einstein's equations, Greek letters like ρ and π are often used as pronumerals, too. The numerals we use – 1, 2, 3 and so on, which have become standard in most parts of the modern world – are Hindu–Arabic in origin, and the other symbols have evolved over time through the work of mathematicians from diverse backgrounds. For instance, the addition and subtraction symbols '+' and '–' were first used in print in 1489 by the German scholar, Johann Widman.

It is important to note, though, that while all peoples have developed some form of mathematics, only a few cultures have significantly influenced the historical development of mathematics as we know it today, and this is also apparent in the symbolism of modern mathematics. Being a universal language – a language of nature and of the human mind rather than of a specific culture – mathematics has needed a cross-fertilisation of ideas from different times and places to reach its current high level of sophistication. But some apparently independent and relatively advanced ancient mathematical peoples historians now know of were too isolated at the time to influence, or be influenced by, the mathematical mainstream. And some peoples did not have the luxury of developing mathematics to such a high degree because of economic hardship or political instability. To give a simplified overview, the main players in the history of mathematics were the Babylonians, Egyptians and Greeks of ancient times, the Indians, Chinese and Arabs of the middle period, and the post-Renaissance Europeans. Nowadays the development of the mathematical language is a global activity. People are still fascinated by this language of pattern for its own sake, as well as for its technological importance.

The numerical patterns in the multiplication tables are expressed in terms of the simple symbols of arithmetic. The grammatical pattern in Newton's and Coulomb's laws is expressed algebraically, but although the symbols express the generality of the pattern, they still represent ordinary numbers or magnitudes. However, according to Faraday, nature contained more complex patterns than that underlying these simple 'Newtonian' forces; fortunately for the theory of electromagnetism, so did mathematics. Thanks to the work of 'pure mathematicians', who study mathematics purely as a language, exploring its symbolic and grammatical patterns simply because patterns are so important to us, so intrinsically satisfying, there exist more complex branches of mathematics than simple algebra, and more complex mathematical entities than simple numbers. Calculus is one such branch, which Newton used when he needed to define the *concept* of acceleration (as a rate of change of speed over time) rather than its numerical *magnitude*. As Maxwell showed, a relatively new branch of calculus, and some new mathematical entities called 'vectors', turned out to be perfect for describing Faraday's field idea, and for summarising the new science of electromagnetism itself. But even this complex new language had been spun from the simple patterns of arithmetic, in a historical and intellectual process of amazing scope and sophistication.

Pure mathematical language

Pure mathematicians begin by analysing the basic language of arithmetic, and they use algebra to explore and generalise its grammatical rules. For instance, if you take the numbers 1, 2, 3, 4 ... – defined to be the 'counting numbers' or 'natural numbers' (because mathematical language includes words as well as numbers and other symbols) – you know from the idea of addition that $1 + 2 = 3 = 2 + 1$. Similarly, $4 + 5 = 5 + 4$, and so on, for any pair of natural numbers. This is one of the rules of arithmetic: you can add numbers in any order.

This rule can be generalised, as a law of addition, and written algebraically as $x + y = y + x$, for any natural numbers 'x' and 'y'. It works for multiplication, too: $xy = yx$. There is a symbolic pattern here, quite aesthetic in an elementary way, called 'commutation', which represents some of the underlying structure of addition and multiplication. It is a 'grammatical' rule of algebra, setting out the allowed relationships between numbers just as English grammar sets out the relationships between words, in the ordering 'subject verb object', for example, or the rule, 'an adjective precedes the noun it describes' (in contrast to French, say, whose grammar generally requires the adjective to be placed after the noun).

Commutation is an example not only of a mathematical pattern, but also of the way ordinary words might have an unfamiliar spin when used mathematically. The familiar use of the word 'commute' is to describe the act of making a regular trip to and from a particular destination. It is used here to describe the *visual* 'to-and-fro' structure of addition and multiplication: in the commutative equations above, 'x' appears at the beginning and the end of the equation, just as a commuter's home appears at the beginning and end of a commutative journey.

The simple commutation pattern works for numbers of

any kind. In *Leaning Towards Infinity*, Sue Woolfe's novel about the mother and daughter mathematicians, Juanita and Frances, Frances discovers a 'new number'. While this idea is left dangling tantalisingly in the novel, it is certainly true that there are different *kinds* of numbers. For a start, there are different kinds of natural numbers. One type that occurs in nature in a fundamental way – particularly in gravitational and electromagnetic phenomena, as the inverse-square laws show – is the 'square number'. It gets its name by analogy with the area of a square, which is a special kind of rectangle. In an ordinary rectangle, the length of each of the two shorter sides is called the width of the rectangle, while that of the two longer sides gives its length or height. A square is a rectangle whose length equals its width, so that all four sides are the same size. If you imagine a square whose sides are all of length 2 centimetres, then its area – *defined* to be its length multiplied by its width – is 2 centimetres x 2 centimetres = 4 'square centimetres'. Similarly, a square whose sides are 3 centimetres long has an area of 3 x 3 or 9 square centimetres, a square whose sides are 4 centimetres long has an area of 4 x 4 or 16 square centimetres, and so on. (And a square whose sides are 1 centimetre long has an area of 1 x 1 or 1 square centimetre.)

As in the inverse-square law, in which r x r is written as r^2, the shorthand notation for the squaring process is 1 x 1 = 1^2, 2 x 2 = 2^2, 3 x 3 = 3^2, 4 x 4 = 4^2, and so on, so that in general, n x n equals n^2 (pronounced 'n squared'), for any natural number 'n' you choose. (A lot of people are confused by this notation, thinking the superscript 2 means 'multiply the number "n" by 2' instead of squaring it. But there is a difference between 2n, the symbol for multiplying 'n' by 2, and n^2, the symbol for multiplying 'n' by 'n'.) All the numbers produced by this process of squaring natural numbers – that is, 1, 4, 9, 16, 25 and so on – are called 'square numbers'.

In a reciprocal definition, the natural number 'n' is called the '(positive) square root' of the square number n^2, because it is the root or origin of the process of producing the square number. So 1 is its own square root (as well as its own square), 2 is the square root of 4, 3 is the square root of 9, and so on. The process of deducing the square root of a square number is called 'taking the square root'; for simple square numbers, this process is so easy you may wonder why it deserves a special name, but finding the square root of 625, for example, involves working out, perhaps by trial and error, that 625 can be factored into 25 x 25. Only then can you deduce that the square root of 625 is 25.

The symbol for a square root is $\sqrt{}$, so $1 = \sqrt{1}$, $2 = \sqrt{4}$, $3 = \sqrt{9}$ ..., but the true nature of the square root operation is easier to see when you write these as $1 = \sqrt{1^2}$, $2 = \sqrt{2^2}$, $3 = \sqrt{3^2}$. You can generalise this pattern, and provide a symbolic *definition* of the square root process, by writing $n = \sqrt{n^2}$. The symbolism shows at a glance that the square root operation 'undoes' the square: you start with 3, say, and square it, to get 9; then take the square root of 9, and you are back where you started, with 3. The square root process is said to be the *inverse* of the squaring one.

Conversely, you can start with the square number 9, and take its square root to get 3; then square this number 3, and end up back where you started, with 9; here, the square is the inverse of the square root operation, and when this sequence is expressed symbolically, you can readily see the way the square 'undoes' the square root: since 3 can be written as $\sqrt{9}$, 3 x 3 = 9 can be written as $\sqrt{9} \times \sqrt{9} = 9$, or, in shorthand, $(\sqrt{9})^2 = 9$. In general, this process can be written as $(\sqrt{n})^2 = n$, for any square number 'n'. (This can also be taken as a definition of a square root: it is a number which generates a square number by the process of 'squaring itself'.)

The idea of an inverse is one of the defining grammatical rules of simple arithmetic. Division and multiplication are inverse operations, and so are addition and subtraction: start with any number, and multiply it by 6, say (or add 6 to it); then divide by 6 (or subtract 6), and you are back where you started. This type of inverse relationship is vital to solving mathematical equations. You find the solution, 'x', to the equation $x + 6 = 10$, by applying the appropriate inverse operation – that is, by subtracting 6 from both sides of the equation (you can do this because as long as you make exactly the same change to each side of an equation, the equation still holds true): $x + 6 - 6 = 10 - 6$, or $x = 4$. Similarly, to solve the equation $x^2 = 4$, apply the inverse (square root) operation to both sides to get $\sqrt{x^2} = \sqrt{4}$, or $x = 2$. The same principle applies to solving more complicated equations, including those involving more complex objects than numbers, and more complex operations than adding and squaring. For the moment, though, I will stick to numbers to illustrate the way the mathematical language works. The way it builds on itself, layer upon layer, in a logically related, increasingly complex web of definition and pattern.

Just as squares and square roots are defined in terms of each other, so new numbers are defined in terms of their relationship to old ones. For instance, the first really different kind of numbers from the natural ones are their opposites, the 'negative' numbers. These are used in everyday language, such as when the temperature is going to be 2 degrees below zero, or –2 degrees. Or when you talk about profit and loss, a profit being a positive number in your accounts ledger and a loss being a negative one. Negative whole numbers, together with the natural numbers, are called 'integers'. Then there are the familiar 'non-whole' numbers, fractions. Fractions – like half or three-quarters (symbolically written as $^1/_2$ or $^3/_4$) – can be positive or negative; they *include* the integers, which can be written as 'improper' fractions like 2/2 (=1) or 6/3 (=2). Fractions and integers are therefore called 'rational' numbers, because they can be written as 'ratios' of two whole numbers, x/y (not because they are intelligent numbers that are able to reason, as the everyday use of the word 'rational' might suggest!).

Because pure mathematics is concerned primarily with symbolic definitions, there is enormous freedom to keep pushing the boundaries of the defining process. It does not matter what meaning your new definitions have, in terms of everyday words and concepts; all that matters is that the mathematical rules are obeyed. (It is the task of 'applied' mathematicians, including mathematical physicists, economists, biologists and encryption experts, to find real-world applications of such pure mathematical definitions.) Fractions and whole numbers (and their decimal expressions) are generally all that are needed in daily life, but when you push the defining process beyond the rational numbers, you start to see the conceptual power of pure mathematical language, because often new and surprising ideas turn up. These frequently translate into powerful new physical definitions, but to pure mathematicians, the process is fascinating in its own right, simply as a linguistic exercise.

For example, take the definition of the square root, $\sqrt{n}$, of a square number 'n', (that is, $(\sqrt{n})^2 = n$). The square root *generated* the square number, so that $\sqrt{n}$ x $\sqrt{n}$ = n (2 x 2 = 4, and so on). From a purely grammatical point of view, however, this definition can be applied to *any* number 'n', not just a square one – at least, it can if you think only about the rule itself, rather than about the problem of defining the 'square root' of a number that is not 'square'. The definition itself simply means that you can write things like $\sqrt{2}$ x $\sqrt{2}$ = 2 or $\sqrt{3}$ x $\sqrt{3}$ = 3, not just $\sqrt{4}$ x $\sqrt{4}$ = 4 or $\sqrt{9}$ x $\sqrt{9}$ = 9. The interesting part is when you ask what sort of numbers $\sqrt{2}$ and $\sqrt{3}$ actually are. They cannot be whole numbers, because the whole numbers 1, 2, 3 and so on generate the square numbers 1, 4, 9 and so on; 2 and 3 are not in this list of whole-number-generated 'square numbers'.

In fact, $\sqrt{2}$ and $\sqrt{3}$ must be numbers larger than 1 and smaller than 2 – approximately 1.4 and 1.7, perhaps – as you can see from the pattern in the following list:

$$1 \times 1 = 1$$
$$\sqrt{2} \times \sqrt{2} = 2$$
$$\sqrt{3} \times \sqrt{3} = 3$$
$$2 \times 2 = 4$$

Like any finite decimal number, 1.4 and 1.7 are rational numbers (they can be written as 14/10 and 17/10); the surprise is that $\sqrt{2}$ and $\sqrt{3}$ *cannot* be written as ratios of natural numbers. They are therefore called 'irrational' numbers. The strange number π, from the expression for the circumference of a circle, $2\pi r$, is also irrational. Mathematicians can easily justify this definition – that is, they can *prove* that $\sqrt{2}$, $\sqrt{3}$ and π cannot be written as ratios of whole numbers – because the mathematical language, based on interrelated definitions and rules of grammar, is self-consistent; any proposition in this language is considered to be true if it is consistent with these definitions and rules. For example, the proposition that $\sqrt{2}$ is rational can be shown to be *inconsistent*: you cannot make *both* the statements $(\sqrt{2})^2 = 2$ *and* $\sqrt{2} = x/y$ (where 'x' and 'y' are integers); the second one is inconsistent with the first – that is, the statement that $\sqrt{2}$ is rational is contradicted by the very definition of $\sqrt{2}$, and this constitutes a 'proof by contradiction' of the irrationality of $\sqrt{2}$. (See the Appendix for details).

The self-consistent nature of mathematics highlights an essential difference between it and ordinary language. The latter is about meaning while mathematics is about grammar, about structure, form and pattern – which, of course, is why it is so useful for expressing the patterns of nature. Not just the simple Newtonian inverse-square pattern, but also more complex ones like the streamlines in rivers and the lines of force around a magnet. The wave patterns in light beams, oceans, sounds, vibrating strings, pulsating heartbeats. The intertwining of energy and matter, electricity and magnetism. And countless other natural processes which seem to reflect the grammatical patterns of mathematics.

Mathematics can only express the more complex natural processes because there is more to the language than numbers and simple, arithmetic-based grammar (although these remain fundamentally important). In fact, humans' fascination with patterns is perhaps more deeply rooted in geometry than in abstract numerical relationships, judging by the number of cave drawings of spirals and circles and other spatial patterns, and in the geometric designs on ancient pottery and other artefacts. Ancient philosopher–mathematicians like Pythagoras and Plato favoured geometry over algebra for the description of the physical world, because its curves and shapes seemed the most direct way to represent physical phenomena like the arcs of the Sun and stars in their journeys across the sky. Plato saw geometry's perfect shapes as pure forms, compared with which physical forms are but imperfect approximations.

Nowadays, mathematicians often try to have it both ways, philosophically speaking. Usually, we say we see it the other way around from Plato (as his famous student, Aristotle, did). Although we can prove mathematical propositions to be completely true, in terms of the rules of pure mathematical language, when we *apply* our equations to describe our sensory impressions of the physical world, we cannot be so sure of their 'truth'. For example, I have pointed out that the Newtonian descriptions of static gravitational, electric and magnetic forces are astoundingly accurate, but they do not fit *perfectly* with the actual physical reality. So mathematicians generally regard the mathematical descriptions as the imperfect approximations; after all, any description, including a mathematical one, can only be an approximation of the thing itself.

But sometimes, when our senses are inadequate for fully perceiving a physical phenomenon, some of us moderns like to think we see true reality in our mathematics; like Plato, we think it is our senses that provide the approximations of reality. As Einstein's $E = mc^2$ illustrates, mathematics is not bound by limitations of physical perception, but only by rules of grammar which, it so often turns out, have an uncanny way of revealing hidden physical relationships. By seeming to favour this latter viewpoint in his theory of electromagnetism, Maxwell was in direct conflict with the philosophy of his day.

His 'Platonic' leanings are also evident in the fact that

geometry was the mathematical language that came most naturally to him; it was also the only mathematics Faraday possessed, and he made good use of it with his pictorial 'lines of force'. Not surprisingly, then, Maxwell found in geometry an important starting point for the mathematical expression of field theory.

Geometry: Pictures and proofs

The ancient Greeks were the first to codify the study of space and shape into a systematic language of geometry. Among the earliest of their mathematicians were the Pythagoreans (the students and later disciples of Pythagoras), who flourished 2,500 years ago. They formed a mysterious community, a strict, secret elite whose rituals included vegetarianism, spiritual purification, and the communal sharing of property and knowledge. Although Greek law forbade women to attend public meetings, the Pythagoreans were more open-minded on this issue, and women like Theano and Perictione were also involved in their mathematical activity. (Theano must have enjoyed mathematics, because she wrote, 'Love is for unoccupied souls'. Perictione wrote, 'We are born to contemplate the world'.)

The Pythagoreans are famous for pioneering the mathematical physics of music, having discovered the relationship between the length of a taut string and the note it produces when plucked. Newton pointed out that they also discovered an inverse-square law, which related the tension and length of the string, each of which affects the octave of the note produced when the string is plucked. (He also noted that the Pythagoreans had used their musical discoveries to build a cosmology in which these same relationships were ascribed to the planets, whose various sizes and orbital dis-

tances were imagined to have the same ratios as those between various musical notes and the associated lengths and
tensions of the plucked strings – ratios that produced the
so-called music of the spheres. It was a fanciful conception,
unlike the practical, objective experiments on music, because the Pythagoreans had no idea what the actual planetary measurements really were. (It would take another three
centuries for realistic, experimentally- and mathematically-
based cosmic measurements to be made.) But Newton was
nevertheless moved to wonder, in a more mystical than
logical mood, whether his own cosmological inverse-square
law had been known to the Pythagoreans. Whether gravity
was, in fact, simply a manifestation of the ancient and
beautiful idea of cosmic harmony!)

In the study of geometry, the Pythagoreans are famous for
Pythagoras's theorem about right-angled triangles: 'The
square of the hypotenuse of a right-angled triangle is equal
to the sum of the squares of the two adjacent sides' – a result
which mathematicians of other cultures had already discovered, notably the ancient Babylonians. (Greek mathematics
was originally founded on that of the Egyptians and Babylonians.) But the Pythagoreans' successors handed down a
logical *proof* of this theorem; in fact, in their uniquely systematic study of geometry, the Greeks, notably Plato, developed methods of mathematical proof we still use today –
methods which are suprising, given their largely geometric
origin, because they are based on *language* rather than geometrical *drawing*.

Perfected and codified by Euclid, around 300 BC, this
linguistic method began, naturally enough, with definitions.
You had to define what you meant by points, lines and
circles. Then there were axioms which were taken to be

'self-evident', like the famous parallel postulate, 'no two parallel lines ever meet (when drawn on a flat surface)'. Then theorems or hypotheses were proposed about less obvious geometrical properties (such as whether or not you can perpendicularly bisect all three sides of a triangle from a single central point). Finally, the pièce de résistance: the proof – or disproof – of the proposition, using only the allowed definitions and axioms. (The number $\sqrt{2}$ was shown to be an irrational number in this way, using the definitions and axioms of algebra.)

The idea of proof has completely changed the way we think, philosophically, scientifically, medically and legally; but only in pure mathematics can it be achieved fully – beyond any doubt, not just beyond reasonable doubt – because of the self-consistency of mathematical language. The mathematical method of deductive proof has come down to us in Euclid's textbook, the *Elements* — the most 'fabulously successful' textbook of all time, having gone through over 1000 editions, and having been used in schools for two millennia, until well into the twentieth century. Today, proofs of geometric propositions are taught in Euclid's style, but with more modern terminology.

For example, an angle is made when two line segments or straight edges meet at a point (in some of the following diagrams, I have enlarged the points so you can easily see

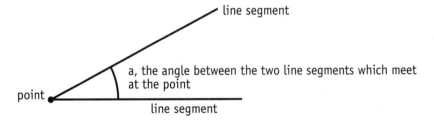

them). An angle is a measure (in 'degrees') of the separation of the lines at this point. It can therefore be described numerically, in terms of the number of degrees it contains. But if you just want to represent an arbitrary angle, you can denote its size by a letter – say 'a' (for angle).

Three important angles are given special names: right angles, straight angles and circular angles. If you rotate, in an anticlockwise direction, the upper line segment shown above, until it is vertical, the angle between it and the horizontal segment is called a 'right angle'; it is defined to have a measure of 90 degrees. (In other words, a degree is one-ninetieth of a right angle.)

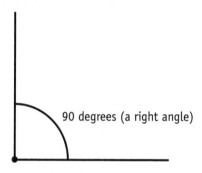

90 degrees (a right angle)

(The term 'right angle' refers to the rectangular shape of the angle, not to the fact that, in the above diagram, it happens to be facing the right-hand side. Nor do the perpendicular lines making the right angle have to be horizontal and vertical – they can both be tilted sideways.)

A 'straight angle' is made by two lines which together lie in a straight line; in other words, if you took the right angle shown above and rotated the vertical arm downwards (anti-clockwise) until it, too, was horizontal, you would have made a straight angle (sometimes just called a straight line).

A straight angle is defined to have a measure of 180 degrees because it contains two right angles.

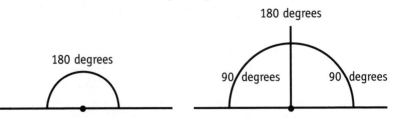

If you then rotate the left-hand arm of the straight angle by another 180 degrees in the anticlockwise direction, until it meets up with the right-hand arm – like clock hands meeting when the time is a quarter past three – you have come 'full circle', and have made a circular angle, which is defined to contain 360 degrees.

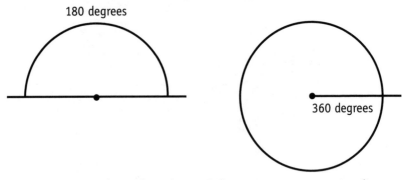

Once you have these basic definitions, you can generalise them. For example, while a straight line can be 'made' from two right angles, it can also be made from any two angles whose sizes add up to 180 degrees.

Angles that lie next to each other like this are called 'adjacent angles'. But you can have more than two angles on a straight line, as long as they all add up to 180 degrees:

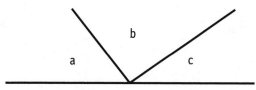

Similarly, you can make up a 360-degree angle from any number of angles – like slices of a pie – but keeping to four 'slices' made by two lines for convenience, 360 degrees can be made from four right angles, like this:

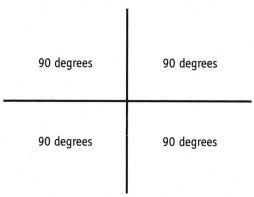

or from any four angles, like this:

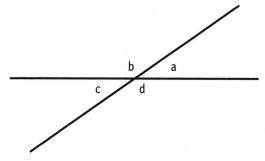

The angles a, b, c and d are completely arbitrary here – I could have drawn them like this:

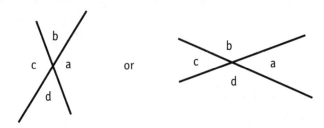

or a variety of other configurations; but using the definition of a straight angle, it is possible to prove some general results about the spatial patterns these angles make (rather than about their actual numerical values). For instance, in each one of these three arbitrary diagrams, all the pairs of angles which lie adjacent to each other on the same straight line must add up to 180 degrees in size: 'b' and 'a', 'a' and 'd', 'd' and 'c', and 'c' and 'b'. This gives a number of algebraic equations for each diagram: b + a = 180, and so on, which you can then use to prove that angles that are 'opposite' rather than adjacent to each other – like 'a' and 'c', or 'b' and 'd' – are equal in size. (Technically, these angles are said to be 'vertically opposite' each other.)

Of course, they *look* equal in the diagrams, and you could check this by measuring them with a protractor, but you can deduce from the definitions that this result is always true, not just for any particular cases you are able to measure. From the definition of a straight angle, you already know that (for any particular diagram) b + a = 180, and b + c = 180. You can then deduce that a = c. (If 'b' were 135, say, you would have 135 + a = 180, and 135 + c = 180, so 'a' and 'c' must each be 45.)

Because all these angles are arbitrary, this result does not depend on any particular configuration; it is a completely general result: vertically opposite angles are *always* equal. So 'b' and 'd' are also equal – you do not have to bother proving it (although you could do so in exactly the same way) because you have already proved a prototype, *general* case. This algebraic facility for generalisation is part of what makes mathematics such an economical language.

Extending the main diagram of vertically opposite angles by adding a second horizontal line, you can create more spatial patterns: the angles 'c' and 'e' in the diagram below lie on alternate sides of the diagonal line, and are called 'alternate angles'. (So are 'd' and 'f'.) The angles 'b' and 'f' lie on the same side of the diagonal line and on corresponding sides of the respective horizontal lines, and so they are called 'corresponding angles'. (Similarly, 'c' and 'g' are corresponding angles, as are 'a' and 'e', and 'd' and 'h'.)

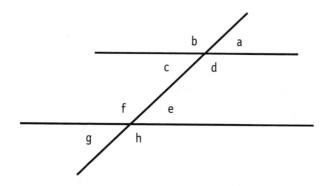

Pairs of corresponding angles are always equal because they are duplicates of each other. (Imagine sliding the bottom horizontal line up until it covers the top one (because they are parallel); then the angles 'e', 'f', 'g' and 'h' fit exactly over the angles 'a', 'b', 'c' and 'd' respectively.) Pairs of alternate angles are also equal, just as pairs of vertically opposite angles are, and you can prove it in a similar way: a = c (because they are a pair of vertically opposite angles); but also a = e, because 'a' and 'e' are a pair of corresponding angles. Therefore c = e. This is a general result, so all pairs of alternate angles are equal in size.

As recorded by Euclid, the Greeks took basic results like these and used them to prove new geometric results, new geometric definitions, and so the language of geometry is built up logically from a few simple concepts about angles and straight lines. For example, the fact that corresponding angles are equal, as are alternate angles, can be used to prove that the sizes of the three angles made by the corners of a triangle must always add up to 180 degrees. This is not intuitively obvious; you cannot tell just by looking at the angles 'a', 'b' and 'c' in the triangle below that their sum is 180 degrees – in contrast to the results about corresponding, adjacent and vertically opposite angles, which *looked* equal in the diagrams. But you can prove that a + b + c = 180 by using the results about alternate and corresponding angles.

In the following diagram, the two lines with the arrow heads are parallel to each other. Angles 'b' and 'd' are equal because they are alternate; angles 'a' and 'e' are equal because they are corresponding; therefore c + d + e = c + b + a, the

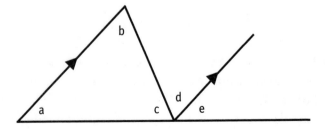

latter sum being the sum of the angles in the triangle. But the former is the sum of the angles on a straight line, that is, 180 degrees, so the triangle's angles add up to 180 degrees too. This result is true for the arbitrary triangle shown, so it is true for any triangle, and becomes a new (proven) geometrical fact or definition.

However, although the logic and economy of geometry is compelling in its own right, the Greek geometers were interested in more than language for its own sake. Three centuries earlier, the Pythagoreans had relied on intuition in developing their 'music of the spheres' cosmology, but now their descendants used geometrical definitions to deduce amazingly good estimates of the size of the 'cosmos', and of the Earth itself. They worked these things out purely mathematically, with the aid of a few local measurements – they had no need to make a trip around the world in order to *physically* measure its size, its circumference, and they had no astronomical tools like telescopes. Just geometry, and extraordinary ingenuity. The word geometry comes from the Greek words 'geo', meaning 'the Earth', and 'metreo', meaning 'to measure': 'geometry' means 'measuring the Earth' – not just its size, but its local distances and shapes, too, although much of the measuring was by deduction rather than tape measure.

For instance, in about 230 BC, a mathematician called Eratosthenes used the geometry of 'alternate angles' to estimate the distance around the circumference of the Earth with an accuracy of almost 95 per cent compared with the modern measurement. His estimate was not completely accurate because of the lack of accurate methods available for measuring angles and long distances, but his mathematics was simple and remarkably powerful. His achievement is even more amazing when you recall that most people at that time believed the Earth was flat.

Calculating the size of the Earth (and proving the world is round)

As viewed from the Earth, the Sun and Moon are circular in shape, as is the Earth's horizon, but while most ancient peoples took it for granted that the Earth was flat, the Greeks assumed all these bodies to be spheres rather than flat circular discs. This was probably a metaphysical conclusion, initially – for instance, Pythagoras considered the sphere to be the most perfect shape – but later Greeks, like Eratosthenes, gave correct mathematical and astronomical arguments for this traditional belief.

It is interesting to digress for a minute and note that many Greek ideas fell into disfavour after the Roman conquest – the Romans being more interested in engineering, administration and law than in pure mathematics or speculative science – and the Romans themselves believed in a flat Earth. It was not until the ninth century AD, when Arabic scholars became interested in ancient Greek knowledge, that the flat Earth idea began to be universally challenged. Through Arabic translations of the Greek works, which were later retranslated into Latin and Spanish by European scholars, ancient Greek ideas were reintroduced into Europe, where they had an extraordinary effect on navigation, cosmology and mathematics itself: at the end of the fifteenth century, the Italian explorer, Christopher Columbus, set off on his epic voyage to America, inspired by ancient Greek 'round Earth' maps (as had been Arabic navigators before him), while in the next century, Copernicus began to develop his heliocentric cosmology after discovering Aristarchus's work, and Kepler referred to the reintroduced work of Apollonius and Archimedes for his understanding of the mathematics of ellipses. Kepler, Galileo and others also used some of Archimedes's and Apollonius's work to lay mathematical foundations for calculus, which eventually came to fruition in the work of Newton and Leibniz. But it was not realised until the twentieth century, when a long-lost Archimedean manuscript was discovered, just how close Archimedes himself had come to the 'integral' branch of seventeenth-century calculus.

When Archimedes was working on his 'calculus', Eratosthenes was the librarian at the great university of Alexandria, in Egypt. (A century earlier, under Alexander the Great – who had been taught by Aristotle – the Hellenic Empire had taken Greek scholarship to much of the known world, from Egypt to India.) It is fitting that it was Eratosthenes to whom Archimedes first showed his manuscript, because the work of both mathematicians predated later European discoveries by almost two millennia.

The first *direct* proof that the Earth is not flat came in 1522, when the surviving members of an expedition led by the Portuguese navigator, Ferdinand Magellan, completed the first circumnavigation of the globe. (Magellan himself was killed before the end of the three-year journey which, having begun with 270 men, ended with fewer than 20.) Back in 230 BC, Eratosthenes had based his argument simply on the noon shadows cast by the Sun at two different places on the summer solstice.

It was known that at the first place, Syene (now called Aswan, in Egypt), the noon Sun was directly (or vertically) overhead at that time of year, because it shone straight down into a deep vertical well. (Consequently, a vertical shadow stick would produce no shadow.) The second place, Alexandria, is 847 kilometres away in a direction which is approximately due north of Syene; Eratosthenes found that the shadows there showed that the noon Sun was not directly overhead on that day, but made an angle with the vertical line of sight. Using a shadow stick in a hemispherical sundial invented by Aristarchus, he estimated this angle to be one-fiftieth of a circle (the idea of degrees not having yet come into widespread use. We would say this angle is 360/50 or 7.2 degrees). The next three diagrams illustrate his process.

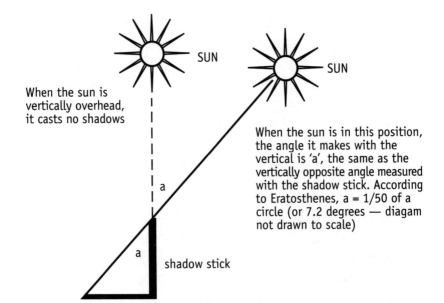

SUN

SUN

When the sun is vertically overhead, it casts no shadows

a

When the sun is in this position, the angle it makes with the vertical is 'a', the same as the vertically opposite angle measured with the shadow stick. According to Eratosthenes, a = 1/50 of a circle (or 7.2 degrees — diagam not drawn to scale)

a

shadow stick

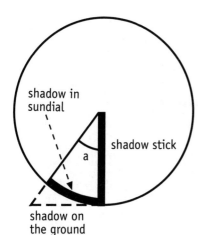

shadow in sundial

shadow stick

a

shadow on the ground

'a' is 1/50 of a circle if the 'slice' shown is 1/50 of the whole 'pie'. In particular, the 'arc' of this slice (or sector) is 1/50 of the circumference of the whole circle. (This is not drawn to scale: 7.2 degrees is quite a small angle.) Eratosthenes measured this arc by letting the shadow fall into a hemispherical sundial rather than onto the ground

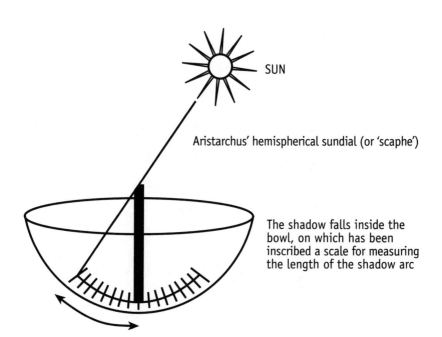

SUN

Aristarchus' hemispherical sundial (or 'scaphe')

The shadow falls inside the bowl, on which has been inscribed a scale for measuring the length of the shadow arc

Armed with this information, and with the belief that the Earth is a sphere, Eratosthenes was able to calculate the circumference of that sphere in an extremely simple way. Taking the Sun to be at the centre of the cosmos, as his older contemporary Aristarchus believed – although the argument can be reversed to apply to a geocentric picture, such as Archimedes supported, and it also applies to an elliptical orbit – this is the way Eratosthenes conceived of the situation: at noon on the summer solstice, Syene and Alexandria are facing the Sun (which thus appears high overhead), and Alexandria is assumed to be due north of Syene, on a meridian of longitude. Each meridian is a complete circle around the globe, the length of the circle therefore being that of the circumference of the Earth.

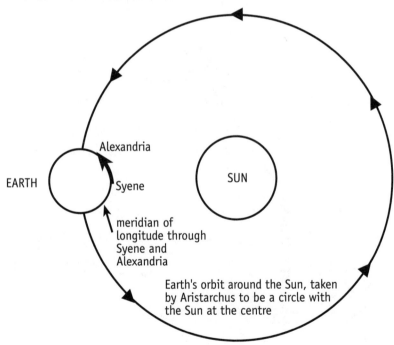

Aristarchus had already made a rough calculation of the relative sizes and distances of the Earth, Sun and Moon, so Eratosthenes was able to assume that, although the Sun is so much bigger than the Earth, it is so far away that it looks like a small ball in the sky (my diagrams are not drawn to scale). This vast separation also means that the rays of sunlight hitting Alexandria and Syene are virtually parallel.

As you can see in the next diagram, the Sun's rays hit Syene directly, and so cast no shadows, but they make an angle at Alexandria; if the Earth were flat, its surface would be a vertical straight line in this diagram (rather than a circle), and the noon Sun would be directly overhead at *both* Syene and Alexandria.

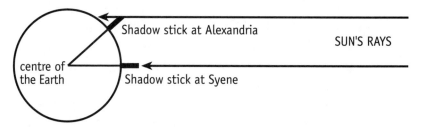

The relevant lines and angles can be extracted from this diagram and represented purely geometrically in terms of alternate angles:

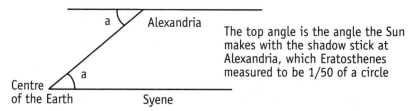

The top angle is the angle the Sun makes with the shadow stick at Alexandria, which Eratosthenes measured to be 1/50 of a circle

Both these angles are equal, being alternate, so the angle at the centre of the Earth, between Syene and Alexandria, must, like the angle of the Sun at Alexandria, be 1/50 of a circle; this means the *distance* from Syene to Alexandria must be 1/50 of the distance around the whole circle from Syene and back again – that is, of the circumference of the Earth.

In this ingeniously simple way, the circumference of the Earth is calculated to be 50 x 847 (= 42,350) kilometres, using the modern measurement of 847 kilometres for the direct distance between Syene and Alexandria. A more accurate measurement of the angle of the Sun at Alexandria – and the knowledge that the Earth is not an exact sphere, but bulges slightly at the equator because of the Earth's rotation, as Newton deduced – would give the modern value of 40,075 kilometres.

Using geometry to find the distance to the Moon

Eratosthenes used simple geometry to imagine the Earth and show just how large it is. Aristarchus had already used some relatively elementary geometry to give an entirely new perspective on the Earth's place in the heavens, by estimating the relative sizes, and distances apart, of the Earth, Sun and Moon. (Newton later used such an estimate of the distance between the Earth and the Moon in his derivation of the inverse square law of gravity.)

Aristarchus used a method of geometrical ratios – involving the radius of the Earth, which he had calculated from the formula for the circumference of a circle, $2\pi r$, using an earlier (less accurate) estimate of the circumference of the Earth and an estimate of the value of π – to calculate *both* the diameters of the Moon and Sun *and* their distances from Earth. You can get a feel for his method by assuming you already know the Moon's diameter, and using the following 'rule of thumb' to find the distance to the Moon. Hold a coin in front of your eye but far enough away so that it just covers the full Moon in the sky.

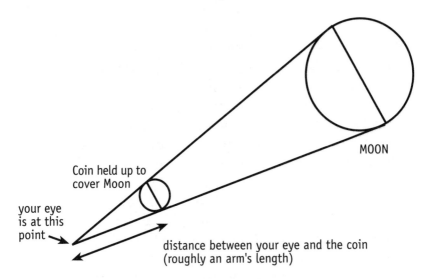

The ratio between the diameters of the Moon and the coin is the same as the ratio between the distances from your eye to the Moon and the coin, respectively (a conclusion which can be deduced with the help of elementary trigonometry, although it looks reasonable if you just look at the diagram). Knowing the diameters of the coin and the Moon, and the distance between your eye and the coin, you can then calculate the Moon's distance from your eye, from the equation

$$\frac{\text{Moon's diameter}}{\text{coin's diameter}} = \frac{\text{distance to the Moon}}{\text{distance to the coin}}$$

Aristarchus's results were nowhere near as accurate as Eratosthenes's, because the measurements at his disposal were inaccurate, but his method was correct; using geometry to provide even a rough idea of the size of the solar system was quite an amazing achievement, and with his method, Aristarchus improved on the geometrical attempts of earlier Greeks to 'measure' the cosmos.

Geometry and algebra unified

Faraday lamented that the only mathematical knowledge he possessed was some basic geometry, but he used his geometrical ideas to imagine aspects of the physical world in an entirely new way, just as Eratosthenes and Aristarchus had done when they put the size of the world in perspective. However, Faraday's simple geometrical conception of his lines of force was not powerful enough to *prove* the physical relevance of fields; it was Maxwell's geometry that enabled the field concept to be expressed in sufficient detail that physical predictions about the nature of fields could be made and tested.

The shape-shifting power of Maxwell's version comes from the fact that it is not based on the kind of geometry expressed solely in terms of ordinary lines, points and angles, but on something far more powerful: algebraic (or 'analytic') geometry. While I have used simple algebra to express the Greeks' 'linguistic' approach to proving geometric propositions about physical lines and angles, analytic geometry can be done *entirely* algebraically: points and lines do not need to be drawn on a page at all, but can be represented by abstract algebraic equations and symbols. Like a poem or story that takes a simple idea and spins it into something altogether more wonderful, 'analytic' language transforms geometry into something which transcends the limitations of our physical senses, and of space itself. It therefore enables physicists to think much more generally and imaginatively about the shape and form of the physical world than they could if they were restricted to concrete geometrical representations of it.

The Greeks started the process of thinking about geometry in terms of systematic language – and they used it to visualise the cosmos in a remarkably modern way. However, it took mathematicians another 2000 years to make the intellectual transition from basic proofs about lines and shapes to a truly transcendent way of imagining the universe.

The long gestation of analytic geometry reflects mathematics's tumultuous history as a universal language. The 'Golden Age' of Greek mathematics came to an end in 212 BC, when Archimedes, deep in the middle of some mathematical calculations, was killed by a Roman soldier.

The Roman conquest certainly had a negative impact on the development of science and mathematics, as I indicated in connection with the lack of interest in (and later loss of) Archimedes's manuscripts and the round Earth idea. However, some historians think the decline of mathematics at that time was due not only to the 'cold breath of Rome', but also to the limitations inherent in the Greek mathematicians' attempts to synthesise algebra and geometry. Perhaps they had gone as far as they could without a cross-fertilisation of ideas from another time or place.

Although they had lost much of their Golden Age tradition, impetus and brilliance, Greek mathematicians continued to work under Roman rule for another 700 years, even producing what is known as a 'Silver Age' from about 250 to 350 AD. The most important mathematicians of this period were Diophantus, who was an algebraist, and Pappus, a geometer who touched on the idea of analytic geometry but then, because he was not really interested in algebra, left it alone. It would take Descartes, a mathematician who was equally interested in geometry and algebra, to successfully unify these two branches of mathematics. Meanwhile, much had still to be done in order for the algebraic language to become sophisticated enough for the task.

The end of the 'Greek' era came in 529 AD, when the Byzantine emperor Justinian closed down the 'pagan' Greek-speaking philosophical academies. Many of the Greek scholars fled to Persia, Syria and elsewhere in the Middle East, while in Europe, 'the spirit of mathematics languished', as the modern historian, Carl Boyer, puts it, 'while men argued less about the value of geometry and more about the way to salvation'. China, India and the Middle East became the seats of mathematical progress.

These regions each had an ancient mathematical history, but they, too, had been through periods in which interest in mathematics 'languished' for various political, religious and cultural reasons. By the end of the twelfth century, India experienced another relative decline in mathematical activity, but had produced many excellent mathematicians in the preceding era, two of the most significant being Brahmagupta, from the seventh century, and Bhaskara, from the twelfth century. Brahmagupta's work includes the first systematic use of negative numbers and zero, while Bhaskara took the first tentative steps in developing the mathematical idea of infinity (as it arises from a division by zero). Mathematicians of earlier cultures had had an idea of the concept of zero, but it was the Indians who developed the arithmetic of zero, treating it as a number in its own right.

The Arabic interest in mathematics also declined around the end of the twelfth century; among the many excellent mathematicians who had arisen during the earlier flowering of Islamic scholarship was the 'father of algebra', al-Khwarizmi, who, in about 840 AD, had written the book whose title gave us the word 'algebra' (*al-jabr*). In this book, al-Khwarizmi did much to put algebra into a systematic framework, just as Euclid had done for geometry. But the mathematician of this era whose name is most well known in the West today – because he was also a poet – is the twelfth-century Persian, Omar Khayyam. He and his colleagues took an important step towards analytic geometry by treating algebra and geometry as equivalent in certain types of problems; a very simple, illustrative example is Pythagoras's theorem, which can be viewed as an algebraic formula for the relationship between the lengths of the sides of a triangle (say, $a^2 + b^2 = c^2$, where 'c' is the length of the hypotenuse, and 'a' and 'b' are the lengths of the other two sides), or as a geometrical drawing of a triangle whose sides are physically drawn so their lengths have this relationship. For the Greeks, algebra and geometry were seen as separate fields, so that when algebraic symbols were used in Greek geometry, they remained philosophically subservient to the geometrical construction itself. The Arabic mathematicians began the process of shifting the emphasis in such situations from geometry to algebra.

Chinese mathematics was still flourishing in the thirteenth century and beyond, when mathematicians like Ch'in Chiu-shao, Chu Shih-chieh and Yang Hui were working on algebra, in particular. Ancient Chinese mathematicians had collaborated with their colleagues from Japan and Korea, and in the early years of the first millennium AD, they had also interacted with scholars from India and Europe, but during the mediaeval era, there were periods in which they tended to be relatively isolated, so that much of their mathematics appears to have been independent of that in other mathematical centres. Certainly they produced some remarkable results which remained unknown in the historical mainstream for centuries. For example, in 1303, Chu Shih-chieh wrote an important treatise on algebra, in which he presented a number of algebraic techniques which, in the West, now bear the names of the Europeans who independently rediscovered them, several centuries later; in particular, the so-called Horner's method, and Pascal's triangle.

As the Greeks had shown, however, even the most dedicated mathematical cultures eventually need fresh inputs, and by the fifteenth century, a reawakened Europe – inspired by the reinvigoration and extension of its Greek heritage through the work of the Arabic mathematicians, and also by the general cultural Renaissance which had begun with Italy's reclamation of its classical Roman roots – had caught up to China and was preparing to take centre stage in the history of mathematics.

The rebirth of European mathematics had begun in the twelfth century, when scholars had learned to translate the Arabic works (which also introduced to Europe some of the Indian mathematics, including that of its systematic number system. This is why we call our numerals 'Hindu–Arabic'). By the early thirteenth century, many European universities had been established, including those at Cambridge, Oxford, Naples and Paris (the Sorbonne). But it took another 500 years, and the contributions of mathematicians from many different European countries (including the Italian priest and professor, Bonaventura Cavalieri, and the amazing French amateur mathematician, Pierre de Fermat), before modern analytic geometry finally began to emerge in the work of Descartes and his successors.

It had been a long and convoluted journey, but at last, through the magical, multicultural synthesis of algebra and geometry, mathematics was ready to become a truly sophisticated language of the universe.

THE MAGICAL SYNTHESIS OF ALGEBRA AND GEOMETRY

'Analytic' or 'algebraically expressed' geometry is fundamental to modern theoretical physics, because of its ability to take the imagination way beyond everyday physical constraints. Newton used an early form of it (in his calculus) to visualise aspects of the mechanism that keeps 'the stars in their courses'; Maxwell used it to imagine Faraday's invisible fields; and Einstein used it to imagine the whole cosmos.

Like Greek geometry, analytic geometry begins with definitions – in particular, the definition of a point. A single point can be drawn with a very sharp pencil on a sheet of paper, or by drawing two intersecting lines or curves, the point being the place where they meet:

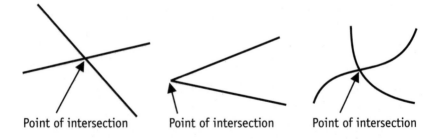

Point of intersection   Point of intersection   Point of intersection

A point can also be defined in terms of its *location* on a page. This is a familiar idea nowadays – it is what you do when you locate a place on a map: its position is described with respect to a grid of horizontal and vertical lines. On a map of the world, these are the lines of latitude and longitude. The Greeks appear to have been the first to come up with this idea.

To define a point or location in terms of a grid of intersecting lines, you have to have reference lines, in terms of which all the other lines are located; the point of intersection of these reference lines is called the 'origin' of the grid. For example, lines of latitude are defined with respect to the equator – the equator is said to have a zero latitude, and all other lines of latitude are measured (angularly) from it. The line of zero longitude is taken, for historical reasons, as the north–south line that goes through Greenwich in England.

On a globe of the Earth, the lines of latitude and longitude are actually circles, but to describe points on a flat surface like a page, a simpler grid can be used. Two perpendicular lines, ruled on the page, are taken to be the zero lines or 'axes' (the bold lines in the following diagram); their point of intersection is the origin, and all the other grid lines are defined in terms of their measured distances from the relevant axis. Graph paper gives a ready-made grid of this type. The distances between grid lines can be measured in any units – millimetres, centimetres, five-millimetre units, for example – depending on the scale needed for the geometrical purpose at hand, but in each grid, the scale has to be consistent. And while geographers define the directions above and below the equator as north and south, this type of grid uses positive and negative numbers to locate lines above and below the horizontal axis; and instead of being east or west of Greenwich, lines to the right of the vertical axis are in the positive direction, while those to the left are negative.

Often, mathematicians only bother to rule in the axes,

but a grid of intersecting lines is imagined to underlie the space defined by these zero lines. Points are represented by pairs of numbers representing their horizontal and vertical distances (in that order) as measured from the axes. The numbers representing the point in this way are called its 'coordinates'. The origin has coordinates (0,0), while the enlarged point in the following diagram has coordinates (3,4): starting at the origin, go 3 centimetres along the horizontal axis; then move up 4 centimetres in the vertical direction to locate the point in question.

Coordinates enable mathematicians to *name* various points – to talk about them as well as to draw them; in analytic geometry, a point is usually thought of as though it *is* its coordinates, rather than a physical dot on a page. (The ancient Greeks were never so bold – not even Apollonius, who pioneered the mathematical use of coordinates; for them, the physical dot on the page was always the important thing.)

The advantage of using symbolic language to represent a

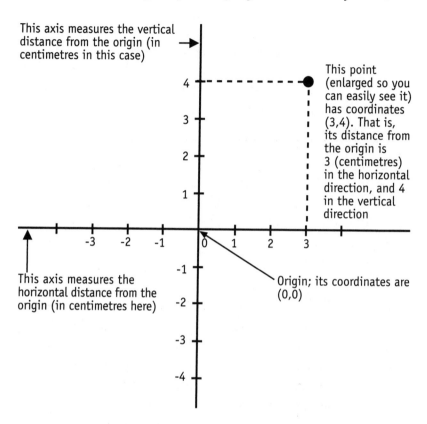

This axis measures the vertical distance from the origin (in centimetres in this case)

This point (enlarged so you can easily see it) has coordinates (3,4). That is, its distance from the origin is 3 (centimetres) in the horizontal direction, and 4 in the vertical direction

This axis measures the horizontal distance from the origin (in centimetres here)

Origin; its coordinates are (0,0)

point – that is, using a pair of numbers rather than a concrete drawing – is that it is easy to make generalisations in this way. To go beyond specific numbers, and into the language of algebra, which enables mathematicians to think both more economically and more broadly. After all, there is not much point, conceptually speaking, in devising an invisible grid just so you can replace a drawing of each individual dot by a pair of coordinates; there is no economising of thought merely in replacing one thing by another. The purpose of the symbolic language is that it enables you to talk about an *arbitrary* or general point. About the very *idea* of a point. An idea that lies at the heart of geometry, because all lines, straight or curved – and all the geometrical shapes that can be made from them – can be described in terms of the points they contain. A line, even a curved one like a circle or spiral, can be described, simply and economically, in terms of a *single*, arbitrary point – there is no need to draw the line, or to list *all* the individual points that lie on it. In this sense, algebra is like poetry. Both get to the essence of things, in an elegant, economical way.

An arbitrary point is represented by arbitrary coordinates, 'x' and 'y' (the horizontal axis is therefore referred to as the 'x axis', and the vertical one is the 'y axis'):

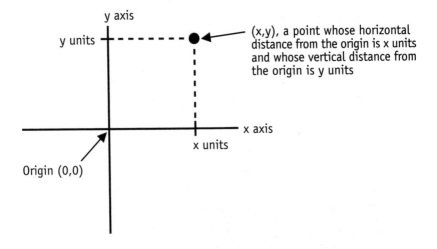

The reason it is possible to describe a geometrical line in terms of a single point is that all the points on a particular line are related to each other by a particular pattern; if they were not, you would just have a random mess of dots, not an identifiable line. Thanks largely to the preliminary work of Apollonius around 250 BC, and to developments by the seventeenth-century mathematicians, Leibniz and Descartes, mathematicians can describe the geometrical pattern of points on a given line with an algebraic equation, which expresses this pattern as a relationship between the coordinates 'x' and 'y' of an arbitrary point on the line.

For example, a straight line drawn through the origin so that it makes a 45-degree angle with both the positive 'x' and 'y' axes (and also with the negative ones), looks like this:

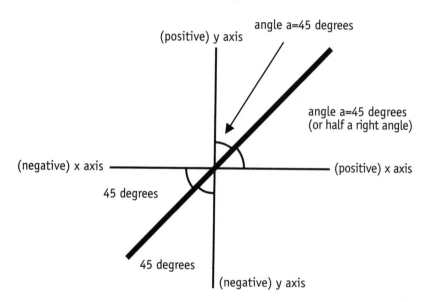

I have drawn this line by eye, using a basic computer package, but to draw it exactly, by hand, I would need a ruler to draw the lines and a protractor to measure the angles. To *describe* the line exactly, using algebraic language rather than a geometrical construction, I need to discover the pattern relating the 'x' and 'y' coordinates of an arbitrary point on it, and I can do this simply by thinking, without making actual measurements. The positive 'x' and 'y' axes make a right angle, so 45 degrees is half way between the axes. The pattern underlying the 45-degree line is therefore very simple: it is a symmetry in which any point on the line is *the same distance from the x axis as it is from the y axis*; this means that its 'x' and 'y' coordinates must be equal, as with the two arbitrarily chosen points highlighted below:

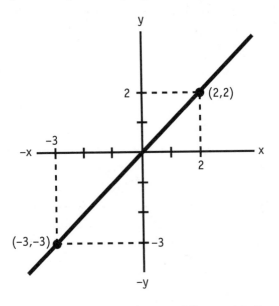

While the points (2,2) and (–3,–3) lie on this line, if you were to draw a line from the origin to the point (1,2), by contrast, or to the point (–2,4) or any point whose coordinates are not equal in size, you would see that it does not make equal (45-degree) angles with the 'x' and 'y' axes. So the 45-degree line shown can be described by the beautifully simple algebraic equation, $y = x$, which expresses the geometrical symmetry of the line in an algebraic relationship between the coordinates (x,y) of each point on the line.

(Similarly, the line y = –x is the 45-degree line through the origin which slopes the other way, in a mirror image of the y = x line. Its points have coordinates that are the same size but with opposite signs, like (–3,3), (2,–2).)

Most lines are more complicated than this, of course, and so are their equations. For instance, the equations for a semicircle (like a rainbow), a parabola (the arc of a water jet, or the path of a ball thrown obliquely into the air), and a catenary (made by a hanging chain) are $y = \sqrt{(r^2 - x^2)}$, $y = -ax^2 + bx + c$ and $y = a\cosh(x/a)$, respectively. The mathematics of these equations is not important here; the point is that visually they look much more complicated than y = x, just as the curves they represent are geometrically more complicated than a straight line.

The advantage of this algebraic type of geometry is that you can conceptualise a line or curve *precisely* – enabling you to differentiate the shape of a rainbow from that of a water jet – and *abstractly*. For example, a straight line is technically infinite in length, so any physical, geometrical representation of it can only be an approximation. The diagram above was large enough to include the points (–3,–3) and (2,2), but from the algebraic equation y = x, you also know that the points (–10,–10) and (99,99) lie on this line. As does the point (0.01, 0.01), which is so close to the origin that it would be impossible to distinguish it from the zero point on the graph drawn above. You do not really need to draw the line, because you can think of it entirely in terms of its equation.

A more important advantage of abstraction becomes obvious when you look at three-dimensional shapes rather than one-dimensional lines (or the two-dimensional shapes made from them). Any two-dimensional shape – whose two dimensions are its length and width – can be drawn with lines on a two-dimensional sheet of paper. For example, the rectangle shown below has its width aligned with the horizontal 'x' axis of the two-dimensional grid, and its length aligned with the vertical 'y' axis. (The axes therefore define the dimensions of a two-dimensional space.)

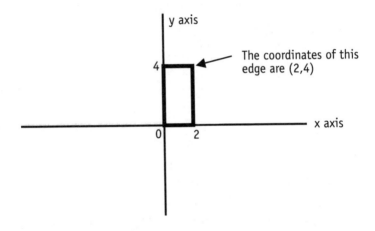

The rectangle is made up of four line segments, so it can be algebraically described by a set of four equations, each describing the coordinate pattern of an arbitrary point on one of the line segments.

A *three*-dimensional shape has thickness as well as length and width; imagine turning the rectangle into a box by adding thickness in the direction sticking out of the page and pointing towards you. It has to be modelled rather than drawn (or else approximated using perspective to give an optical illusion of three dimensionality) because drawing three-dimensional images on a two-dimensional page is technically impossible. But you can easily represent a generic point in such a space in terms of its coordinates: while a point in two-dimensional space has two coordinates (x,y), a point in three-dimensional space is represented by three coordinates, (x,y,z), all measured from a specified origin, (0,0,0); the 'z' axis is perpendicular to the plane of the page on which the 'x' and 'y' axes are drawn. Lines and shapes in three-dimensional space can then be described algebraically, by equations relating the underlying patterns between the three coordinates of relevant arbitrary points.

Four-dimensional geometry

The transcendent power of the symbolism reveals itself, however, when you mentally combine the three *spatial* dimensions with *time*. As Einstein showed (building on the ideas of his former teacher, Russian-born Hermann Minkowski), you can then create a whole new vision of the physical world, based on the four-dimensional geometry of 'spacetime'.

Geometry is about points and lines, and the shapes made from them, but it is impossible for most of us to visualise four-dimensional versions of these objects because we live in a three-dimensional world. It is therefore impossible to draw or construct a four-dimensional point or line. Such things only exist in the algebraic language: a point in spacetime is described analytically by *four* coordinates (x,y,z,t); written like that, a four-dimensional point is easy to conceptualise. The rest of the geometry can be built out of these points.

Einstein imagined that empty spacetime is flat, but that the presence of matter curves the spacetime around it, just as a heavy bowling ball placed on a flexible flat surface like a trampoline will make a dent, bending the surface around it. The basic description of flat spacetime is contained in Einstein's special theory of relativity. This theory was conceived to explain the influence of motion on perception, not to describe spacetime, and it was first published, in 1905, in a purely algebraic form having nothing to do with four-dimensional geometry. It was Minkowski who showed, in 1908, that special relativity could be understood more simply by imagining the motion taking place in four-dimensional spacetime, rather than *in* three-dimensional space *over* time. (It was also Minkowski who had once called Einstein a 'lazy dog', because as a student he hardly ever bothered to turn up to lectures.)

Minkowski's metaphor immediately took on an aura of physical reality, because the idea of spacetime implied that space and time are physically intertwined in some fundamental way, and this seemed to explain special relativity's counter-intuitive assertion that one's perception of *time* depends on one's motion through *space*. Most of us assume that (accurate) clocks keep steadily ticking away independently of our own activities, so that time passes at the same rate for everyone. But suppose you leave the Earth in a very fast space ship, journey through space for seven years, and then head back to Earth; the equations of special relativity predict that on returning home, you will find that everyone will have aged not fourteen years, as you will have done, but fifty. According to your Earth-bound friends, time really will have slowed down for you, although you would not have noticed anything temporally unusual at the time. Such an effect has subsequently been observed in the laboratory, using not fast-moving astronauts but fast-moving radioactive particles, which live longer (before their complete radioactive decay) than do identical particles at rest. It has also been shown that time slows down measurably (although barely so) even on a routine aeroplane flight. These are but two of the physical tests that have supported Einstein's mathematics. Mathematics that has completely changed physicists' philosophical notion of reality, because the relativity of time has more profound consequences even than the possibility of time travel. The intertwining of space and time means that at a fundamental level, even the distinction between past, present and future is relative.

Long before special relativity had been experimentally vindicated, Einstein began building on the idea of four-dimensional spacetime, and in 1915 he completed the *general* theory of relativity, a description of the geometry of curved spacetime. In physical terms, it is the *gravity* associated with matter that makes the spacetime curve, so general relativity is an extension of Newton's theory of gravity. The new theory predicted some even more bizarre aspects of the physical universe than did special relativity: that the universe itself must either collapse under its own weight – a possibility Newton had been confronted with in his own theory of gravity – or it must actually be *expanding*. This dual consequence of Einstein's equations was first taken seriously not by Einstein but by the Russian mathematician, Aleksander Friedmann.

The prediction of the eventual collapse of the universe arises in both cases from the fact that all the planets, stars and galaxies are gravitationally attracted to each other – just as an apple and the Earth attract each other; without some other force to counteract gravity, the galaxies would 'fall' into each other, and eventually the whole universe would merge into one central heap of matter. In desperation, Newton had invoked God as the source of this counter-force, a ploy that had provoked a sharp rebuke from Leibniz (who was already alarmed at what he saw as Newton's mechanistic view of nature): 'Surely God doesn't need to *wind up* his watch!'

Laplace later showed that as far as the solar system went, Newton's equations actually predicted a *stable* system (this is what had prompted Napoleon to ask Laplace where God fitted into his analysis). Since the universe as a whole did not seem to be collapsing either, and since God had long been dispensed with as a fix-it man to patch up holes in scientific theories, Einstein had modified his original equations of four-dimensional curved spacetime by introducing an 'anti-gravity' term, called the 'cosmological term', which he hoped would perfectly counterbalance the attractive force of gravity and ensure his equations would not predict a gravitational collapse. But Friedmann's work suggested there was a simpler way to mathematically avoid such a collapse. By choosing the expanding option in his expanding/contracting solution of Einstein's original equations, a model of the universe is created in which the force of its expansion more than counteracts the tendency towards gravitational collapse.

A way to visualise what an expanding universe looks like is to imagine some bread dough in which raisins have been mixed; the dough is quite compact at first, but as it rises, the raisins, which represent stars and galaxies in an expanding universe, move further apart from each other. Or you can imagine two paper stars stuck onto the outside of a balloon; when someone blows up the balloon, the distance between the stars increases as the balloon expands. In other words, as the universe expands, so does space itself. Physically, however, the universe appeared static and unchanging, so Einstein initially paid little attention to Friedmann's analysis.

But others besides Friedmann had more faith in Einstein's mathematics than did Einstein himself. The English astronomer, Sir Arthur Eddington, and the Belgian priest and astrophysicist, Georges-Henri Lemaître, also began to develop expanding cosmological models out of Einstein's equations. (They also showed that the cosmological (anti-gravity) term did not achieve its intended purpose because the balance between gravity and antigravity was too unstable – too exactly finetuned – to be achievable in the real world; in nature, things are never balanced with quite the precision of mathematical equations. Lemaître and Eddington showed that the anti-gravity term would end up producing not a perfect, 'static' balance between expansion and gravitational collapse, but an increasingly expanding universe.)

Then, in 1929, something extraordinary happened. The American astronomer Edwin Hubble found physical evidence which suggested that the universe *is* expanding, evidence based on the now famous red-shift, an observed shifting of the wavelength of starlight towards the red end of the spectrum. It is the same kind of observable shift in wavelength that makes the sound of the siren of a receding ambulance, or the whistle of a departing train, get lower as the vehicle moves away from us, and higher as it approaches – although unlike the sound of a siren, the red-shift can be 'observed' only with technical equipment and know-how. This kind of change in the wavelength of sound and light waves is called a Doppler shift (after the Austrian physicist Christian Doppler): as the sound or light source moves towards you, the distance between you and the source decreases so the wave appears to be 'compressed'

into this smaller space, making its wavelength shorter –
which means its sound is higher or its colour is shifted
towards the shortwave or blue end of the spectrum. When
the source moves away from you, the wave is 'stretched' to
fill the increasing distance between you and the source; the
sound therefore gets lower or the colour moves towards the
red end of the spectrum.

Hubble found that the light from most of the galaxies in
the sky was being red-shifted rather than blue-shifted (or left
unchanging). But if light from the distant stars is red-shifted,
then, like the ambulance or train, these stars must be receding from us, speeding away from us as the space between us
expands.

With Hubble's monumental discovery, Einstein regretted
his lack of mathematical faith, saying it was the biggest
blunder of his life. But the discovery helped validate the
general theory of relativity, revealing it as another example
of the way mathematics can take physicists beyond thought
itself: no one had seriously imagined that the universe is
expanding before it turned up as an unintended consequence of Einstein's equations for the geometry of four-
dimensional spacetime.

That was not the end of it, though. Hubble's discovery
meant that some philosophically uncomfortable questions
had to be asked about the *origin* of the universe. What was
it expanding from? When did it start? Many scientists were
uncomfortable with the idea that there might be a beginning
of time, because what happened before time existed? Others
did not like the religious implications that there might have
been a moment of creation. Conversely, many did not like
the fact that science was once again straying into what they
saw as religious territory.

In 1931, Lemaître formally sowed the seeds of the Big Bang theory, when he suggested the universe had started as an explosion of a 'primeval atom', and that it had continued expanding from that explosive beginning. Some of the world's more ancient creation myths have also imagined the world exploding from some sort of cosmic seed or 'egg', but they have not seriously suggested that it is still expanding, or described it in the kind of extraordinary detail with which mathematical physicists and cosmologists have subsequently described the Big Bang. Which does not necessarily mean the mathematicians have it right, but it does show what an incredibly rich language mathematics is.

For instance, in 1970, the English physicists, Stephen Hawking and Roger Penrose, showed that Einstein's equations predicted the universe had expanded not from a tiny piece of matter located in an otherwise empty cosmos, but from a *single point* in four-dimensional spacetime. This meant that the Big Bang was not an ordinary explosion which took place at a specific three-dimensional location at a given time on a pre-existing cosmic stage, but that space and time themselves were actually created in the explosion, along with all the matter and energy that has evolved into the galaxies, stars and living things that now fill the observable (and much expanded) universe. Before this point, about thirteen billion years ago, there was no time and no space. No geometry, no matter, nothing. The universe simply appeared out of nowhere. Out of nothing.

Such is the supposed creative power of the Big Bang. Some philosophers, notably the early Christian, St Augustine, had considered the possibility of an origin of time, but they could not put so accurate a date on it. Incidentally, 'Big Bang' is the delightfully evocative name coined in 1950 by the English astronomer, Fred Hoyle, in order to *disparage* the idea of a moment of creation. Despite the red-shift evidence of 1929, the debate over the shape and creation of the universe remained open-ended until the 1960s, and even today, some scientists are not entirely convinced about how to interpret the increasing amount of astronomical evidence.

But most physicists agree that physical echoes of the Big Bang have now been detected, notably in the 'cosmic microwave background radiation' discovered in 1964 by the American radioastronomers, Arno Penzias and Robert Wilson. This is a very faint 'bath' of heat that fills the universe. According to the Big Bang theory, in the initial inferno of creation, the temperature would have been incomprehensibly enormous, but as the universe has expanded over billions of years, its temperature should have dropped to about minus 270 degrees centigrade. This barely registers on the temperature scale, but nevertheless, even 'heat' this cold should be detectable as part of the background energy of space – a background energy that should exist, according to the theory, because energy and space were created together in the Big Bang; they are inseparable. The 1964 discovery of just such a bath of 'heat' was powerful evidence in support of the theory.

Physicists immediately realised that this ancient heat radiation must carry clues about what the early universe was like. The heat seemed to be extremely uniform – it had the same temperature everywhere the astronomers looked. But this was a problem for the Big Bang theory – which describes not just the origin of the universe but also its evolution and expansion – because, while the present universe is extraordinarily uniform on a large scale, its finer structure is rather lumpy, its matter existing in clumps (like the galaxies) rather than being spread evenly throughout space. To explain this present lumpiness, the theory assumes there must have been irregularities in the early universe, too – irregularities that should have shown up as temperature variations in the primeval heat bath.

The reasoning is essentially this, based on mathematical, astronomical, physical and chemical considerations which have combined to produce the details of the mathematically inspired Big Bang theory: the dominant force in shaping the universe is gravity; the early universe was so hot that matter only existed in a 'vapourised' state, but large-scale structures like the stars and galaxies eventually formed out of these primordial clouds of gas by a process of 'gravitational accretion'. You can understand this by recalling Newton's law of gravity, which says the force of gravity emanating from an object is proportional to its mass. For instance, the Earth exerts more force on a person on its surface than the Moon does on a lunar astronaut, who falls to the ground more slowly – that is, with less force – than he or she would do on Earth. In the case of the early universe, the 'object' I am talking about is a whole region of gas, so the more gas (or mass) in a given region, the stronger the force of gravity it will exert on the gas in other areas. If all the regions in the early universe had the same density of gas, then no one region would gravitationally influence its neighbours more than any other. If, however, the gaseous matter in some places was denser than in others, so that there were more particles – and therefore more mass – in those areas, then the force of gravity from these patches of gas would be stronger than elsewhere, and each of them would attract matter from other regions. As more gas was pulled into a region, the region's gravitational pull on the gas in surrounding areas would have become even stronger, so it would be able to reach further afield and draw in even more gas. By such a process of gravitational accretion, in which clumps of gas became denser because of gravity, stars and galaxies eventually formed.

According to the Big Bang theory, then, the original variations in the density and gravity of the cosmic clouds effectively created gravitational 'seeds', onto which the galaxies accreted. But these primordial gravitational variations should show up as temperature variations in the ancient background radiation, because energy is also affected by gravity, according to general relativity theory. Einstein gave an elegant 'gravitational red-shift' argument for this, which essentially says that, since gravity causes things to fall – to move – it must produce a Doppler shift in the wavelength (and therefore the temperature) of light and other radiation, just as the relative motion of the stars does. Different amounts of gravity produce different amounts of shifting, and these differences would show up as variations in background temperature across a region of varying gravitational intensity, just as the light from distant stars is red-shifted by different amounts (producing different colours), depending on how quickly they are receding from us.

In 1964, no such temperature fluctuations in the cosmic heat bath were detected, but in 1992, after a painstaking analysis of new measurements made by NASA's Cosmic Background Explorer (COBE) satellite, a team of physicists led by George Smoot triumphantly reported that there are, indeed, infinitesimal irregularities (or 'wrinkles') in the temperature of the primordial background radiation.

The discovery of these cosmic wrinkles was another huge

boost to the Big Bang theory. More recent evidence, from astonomers in Australia and the United States, suggests the universe is not only expanding, but is doing so at an increasing rate, which is good news for the Big Bang expanding-universe theory, and suggests that perhaps Einstein's cosmological term, and Eddington and Lemaître's analysis of its physical importance, was actually valid. (After all, Einstein had not just added the term as though it were a bandaid, propping up his equations in the light of then current experimental evidence. Rather, he had explored the mathematical equations themselves for 'clues', and so his additional term was mathematically consistent, although it did not produce the static cosmological description Einstein had initially hoped for.)

Einstein's four-dimensional mathematics has conjured up a host of other cosmic details, one of the most famous being a description of a 'black hole'. John Michell and his contemporary, Laplace, conceived the basic idea, but Einstein's mathematics provides a detailed description which seems more like a scene from a horror movie than a mathematical analysis. A black hole is a region of spacetime that is so strongly curved by gravity that nothing, not even light, can escape it; its relentless gravitational force sucks in everything in its way and then tears it apart, pulling the pieces down to its mysterious heart, where space and time turn themselves inside out and a new world begins – or ends. In fact, some say the universe itself may be a colossal black hole.

They seem too weird to be true, but black holes are now considered to be a real cosmic phenomenon, evidence of their existence being inferred by the otherwise inexplicable behaviour of nearby stars which are in the early stages of being torn apart, or which seem to be close enough to feel the force of gravity from these invisible powerhouses – so their orbits are distorted – but not so close that they are sucked inside. In the latter case, it is the same kind of inference which led Adams and Leverrier to predict the existence of the then invisible Neptune because of observed 'wobbles' in Uranus's orbit.

No one will ever be able to prove beyond doubt that mathematical creations like black holes exist and that the Big Bang actually happened, because we can never return from an exploratory trip into a black hole, and we can never repeat the experiment of creation! Some physicists, like Hawking, believe that further mathematical analysis, involving quantum mechanics as well as general relativity, might replace the idea of an actual moment or point of creation with something less 'supernatural'; but although these and other such details will no doubt be further refined as more mathematical and experimental information come to light, the basic picture of the Big Bang is not expected to change significantly, given the amount of experimental evidence now available.

However, whatever the ultimate reality of our origins – and whatever the fate of relativity theory itself – physicists have learned an incredible amount about the universe over the past century, guided by Einstein's mathematical theory of gravity. A theory that could not have been conceived if Einstein's imagination had been limited by the kind of geometry produced with a ruler and compass, or otherwise physically constructed in our ordinary three-dimensional world. As a language for describing the physical universe, mathematics has developed its extraordinary power through the synthesis of algebra and geometry into 'analytic geometry' – a synthesis that has enabled physicists to think in entirely new ways about the physical world of space, shape and time.

While Einstein and his followers took the idea to astronomical heights, Maxwell had used analytic geometry to re-imagine the universe on a much subtler scale – that of the electromagnetic field. In doing so, he resolved the field versus action-at-a-distance controversy, prompting Einstein to say, 'The greatest alteration in ... our conception of the structure of physical reality since the foundation of theoretical physics by Newton, originated in the researches of Faraday and Maxwell on electromagnetic phenomena'. In particular, Maxwell. 'Before Maxwell' (as Einstein put it), physicists conceived of physical reality in one particular way – the Newtonian way, in which the world was considered to be made up of separate objects (or particles); after Maxwell, they were able to see reality holistically, in terms of continuous fields – a viewpoint that has proved both philosophically 'profound' and practically 'fruitful'.

Einstein was speaking with the authority of hindsight – the authority of having successfully used Maxwell's theory of electromagnetism as the basis of special relativity, and having used his mathematical field language in general relativity theory. Like most radical changes, however, Maxwell's new theory was initially controversial. He published it in 1865, but even in the 1890s, many action-at-a-distance devotees did not accept it, and as a student in Zurich, Einstein had had to teach himself about it because his teachers did not include it in their curricula. Neither did Maxwell's friends, Stokes and Thomson. But I have jumped way ahead of my story. However, with hindsight we can appreciate that just as Maxwell's genius enabled him to see the significance of Faraday's unpopular field idea, so did Einstein's genius manifest itself in his acceptance of Maxwell's controversial mathematical field theory.

MAXWELL'S MATHEMATICAL LANGUAGE

Setting the scene

Maxwell actually produced two challenges to Newton's legacy: not just the field theory of electromagnetism but also a statistical analysis of gases. The former was a direct assault, but the other was indirect – a fact that helps explain why Maxwell's fame during his own lifetime rested on the latter achievement. However, Maxwell never believed his field theory diminished Newton's achievement; like Faraday, he knew the great man himself had been unhappy with the idea of action-at-a-distance, and he believed he was continuing Newton's work rather than challenging it.

As for his work on gases, it actually arose out of Newtonian ideas, although it laid the foundations of a whole new branch of physics – statistical mechanics – which, half a century later, blossomed into quantum mechanics. At about the same time, the early twentieth century, Einstein created his relativity theories, which also challenged the Newtonian way of looking at reality. Or rather, as in all these cases, they challenged the *range of application* of the Newtonian view. Newton's methods and perspective are still valid in many situations, particulary at the everyday scale of perception. Quantum theory, on the other hand, applies at the infinitesimal, subatomic level, while relativity comes into its own on the vast, cosmic scale, or when things are moving extremely fast, like light. The problem with mid-nineteenth-century Newtonianism was simply that Newton's theory of gravity was then the only satisfactory mathematical theory of physics, so theoretical physicists tried to apply his methods to *all* physical phenomena.

In the case of gases, however, Newtonian concepts provided an excellent starting point. The idea that matter was composed of molecules was only a tentative hypothesis at that time, but the experimentally observed behaviour of gases could be neatly explained by assuming they were composed of fast-moving particles. For example, a gas contained in a jar exerts a pressure on the walls of the jar which can be explained in terms of collisions between the fast-moving molecules and the walls. (This is the kinetic theory of gases, developed over two centuries by Boyle, Bernoulli, Joule, Clausius, Boltzmann and Maxwell, among others.) Incidentally, Maxwell also embraced John Herschel's idea of using molecules, supposedly enduring and unchangeable, as evidence of God's design in nature; Darwin's theory had just been published, and the 'miracle' of biological perfection could no longer be used as such evidence.

At the heart of Maxwell's statistical method was the idea of probability. If you toss a coin 50 times, it is likely you will get about 25 heads and 25 tails. In other words, heads are likely to turn up half the time. In a single toss, therefore, the probability of getting a head is said to be a half, or one chance in two. Maxwell wanted to know the probability of a gas molecule attaining a particular speed, given the measured pressure of the gas on the container walls, and also its temperature and density. (According to the kinetic theory, the more pressure the molecules exert, the faster they must be moving. Also, the higher the temperature of the gas, the faster the molecules are assumed to be moving.)

The use of probabilities to describe physical reality was a paradigm shift; previously, all particles had been treated deterministically, using Newtonian methods: for example, if you throw a ball into the air, you can use Newton's Second Law of Motion (F = ma) to determine where it will be at any later time, and how fast it is moving. However, in a gas which is supposed to be composed of millions of molecules, it is impractical to apply Newton's law to each and every one of them, and so this kind of mathematical control over the whereabouts of gaseous particles had to be relinquished. In the statistical theory of gases – and in the quantum theory of subatomic particles – only the *likely* or *probable* positions or speeds of individual particles are calculated (although in the quantum case, there are other, more fundamental reasons for this than just the practical difficulties in making measurements).

In 1859, Maxwell succeeded in working out a probability equation for the speed of gaseous molecules, given various experimental measurements of the gas's pressure, temperature, and so on. (Because of the difficulty of measuring even a sample of molecular speeds, it took almost a century for his statistical equation to be physically verified, thus providing further support for the kinetic theory.) The following year, he left Aberdeen and took up a professorship at King's College, which was in London, where Faraday lived, and for the first time, he and Faraday met. Maxwell regularly attended Faraday's popular Friday evening lectures, and one night, as a particularly large and jostling crowd was leaving the auditorium, Faraday called out, 'Ho, Maxwell! You cannot get out? If any man can find a way through a crowd it should be you!' He was referring to Maxwell's work on mathematically describing the jostling molecules in gases.

Maxwell spent five extremely productive years at King's College. During this time, he and Katherine lived in Kensington, close to Hyde Park, where they used to ride their horses in the afternoons. The house had a long attic, which Maxwell converted to a laboratory where he performed further experiments on the behaviour of gases at various temperatures, Katherine stoking the fire and regulating the temperature. They also did experiments on colour, which initially terrified the neighbours, who saw Maxwell, sometimes late at night, staring into what appeared to be a coffin; it was actually a homemade 'colour box', which he used in his study of colour vision – a study that led him to produce what seems to have been the first coloured photographic image of an everyday object. He made transparencies from three black-and-white photographs he had taken through three different-coloured filters: red, green and blue; he then simultaneously projected the transparencies, using red, green and blue light, to produce a reasonably accurate image of a three-coloured ribbon. He demonstrated this achievement, which was important for the later development of colour photography, to the Royal Society, in 1861; Faraday, who was also interested in the new science of photography, and who had an extensive private portrait collection, was in the audience.

At this time, Maxwell also produced several papers on electromagetism, including the one in which he presented his controversial theory, *The Dynamics of the Electromagnetic Field*, which was published in 1865. In that year, at 34, he temporarily retired from academic life, saying he was returning to Glenlair to 'stroll in the fields and fraternize with the young frogs and old water-rats'. And to attend to his duties as laird: visiting tenants, caring for them when they were sick (including reading the Scriptures at their bedsides 'where such ministrations were welcome'), and entertaining them when they were well – 'dining the valley in appropriate batches', as he put it.

During these years back at Glenlair – from 1865 to 1871, when he became the first director of Cambridge's first science laboratory, the Cavendish – Maxwell also worked on his most famous and wide-reaching publication, *A Treatise on Electricity and Magnetism* – an attempt at expressing his controversial ideas more accessibly. Writing more than a century after the *Treatise* was first published in 1873, Maxwell's modern biographer, Martin Goldman, has justly called it 'a magnificent book, many sections of which could still be used today (some parts of it are better than today's texts)'. Not that it was polished, or without critics, but the difficulties in the book are not due to Maxwell's writing style, but to the fact that the physics of electromagnetism was then still relatively rudimentary, and confusion sometimes arises from Maxwell's own confusion about some of these physical ideas – and from the fact that some of his choices of mathematical symbolism (which now appear archaic, and so make the *Treatise* difficult in parts for modern readers) were too *new* for many of his contemporaries.

Maxwell's special insight into the fundamentals of scientific communication hinged on his conscious commitment to making scientific ideas accessible, to both scientists and the public. To reach a wider audience, he wrote encyclopaedia articles and book reviews, but even in his own scientific writing, he always tried to enlighten rather than dazzle his readers. (While working on his paper on Saturn's rings, he had written to a friend, 'I am very busy with Saturn, on top of my regular work [as professor at Aberdeen]. He is all remodelled and recast, but I have more to do to him yet, for I wish to redeem the character of mathematicians, and make it intelligible'.)

As a referee of others' scientific papers, he advocated another strategy modern scientists would do well to follow: do not revise papers so much that the original ideas which sparked the discovery are lost in a sophisticated presentation of the final conclusions. Even the great Ampère, he said, had fallen into that trap, while Faraday, on the other hand, was meticulous in presenting the whole process of his experiments. Maxwell compared the two approaches in a delightful passage from the *Treatise* (in which he speaks first of Ampère's account of his experimental extension and mathematical expression of Oersted's discovery).

> The whole, theory and experiment, seems as if it had leaped, full grown and full armed, from the brain of the 'Newton of electricity'. It is perfect in form, and unassailable in accuracy … [but Ampère's presentation] does not allow us to trace the formation of the ideas which guided it …
>
> Faraday, on the other hand, shows us his unsuccessful as well as his successful experiments, and his crude ideas as well as his developed ones, and the reader, however inferior to him in inductive power, feels sympathy even more than admiration, and is tempted to believe that, if he had the opportunity, he too would be a discoverer. Every student should therefore read Ampère's research as a splendid example of scientific style in the statement of a discovery, but he should also study Faraday for the cultivation of a scientific spirit …

Of course, the deeper ramifications of this difference in style between two of the leading protagonists in the electromagnetic controversy did not escape Maxwell, who felt that Faraday's field idea was less polished but more true to the spirit of electromagnetism than were Newtonian approaches like Ampère's. Which was why he had taken on the field idea: he believed it held the key to understanding the true nature of electromagnetism, without which it would be impossible to build a complete mathematical description of the known phenomena (the static electric and magnetic forces, and electromagnetic induction). Besides, he had wanted to create not only a mathematical *description* of the observed facts of electromagnetism – the way Kepler had described the orbits of the planets – but also a theory about why or how they happened, just as Newton had done for astronomy. Faraday had created the basis of field theory, but Maxwell intended to give substance to fields by describing them precisely.

Knowing that the best mathematicians of the day had adopted the Newtonian view of electromagnetism, he must have had an extraordinary belief in himself to challenge them, almost single-handedly, especially in light of his acknowledged mathematical limitations: aside from his famous carelessness, he once wrote to Tait berating himself for his 'invincible stupidity' in having taken 'too long' to solve a problem he had been working on.

Tait later said that technically, Maxwell was not a first-rate mathematician – he was slow in 'writing out' the full implications of symbolic equations, preferring to think in terms of geometrical or physical images, later translating his ideas into the necessary algebraic symbols. However, Tait continued, Maxwell had an 'extraordinary power of penetration', an 'altogether unusual amount of patient determination', and a 'clearness of vision' which enabled him to use mathematical language in an original and imaginative way, and which therefore made him, 'in the true sense of the word, a mathematician of highest order'.

In fact, Maxwell produced some of the most beautiful and practical equations ever written. He achieved this by realising that a new branch of analytic geometry, called 'vector calculus', was the ideal language with which to describe Faraday's field (just as Minkowski later showed that the analytic geometry of four-dimensional spacetime was the ideal language for Einstein's special relativity theory).

Vector fields

As far as the basic static phenomena were concerned, Maxwell had no argument with the Newtonians on what actually happens to one electrically charged particle, or one magnet, when it is placed near another. After all, the experimental science of static electricity and magnetism was two centuries old, and it had long been demonstrated that if you take a positively charged object – such as a glass ball which has been rubbed with silk – and attach it to a table, then when you place on the table, next to the fixed ball, another positively charged ball (called a 'test particle'), it will be repelled away from the fixed ball with a force whose strength can be calculated from Coulomb's law. If the fixed particle is negatively charged, the test particle will be attracted to it. You can create a similar set-up with a pair of long bar magnets, whose separate poles also attract or repel each other according to Coulomb's law. (One member of the pair of particles in each set-up is fixed, because otherwise both of them would move; the force between them is mutual, thanks to the Third Law of action and reaction. But just as it is easier to investigate gravity by considering the Earth to be fixed, so that only the apple moves, so it's easier to investigate electric and magnetic forces by focusing on the motion of only one of the objects.)

The disagreement concerned not what happened but how it happened. In the case of the two charged particles, the Newtonians assumed that the electric force on the test particle acted at a distance, instantaneously jumping from the fixed particle to the test one. They assumed only the particles themselves were affected by this (mutual) force, but Faraday supposed it to affect every point in the space between the particles, via 'lines of force'. He believed the invisible lines of magnetic force around a bar magnet had been made manifest by iron filings. In the electric case, by placing the test charge at various points around the fixed charge, and recording, each time, the path along which it was repelled, Faraday deduced that the fixed particle was surrounded by invisible lines of electric force radiating out from the fixed charge. (If the fixed particle were negatively charged, a posi-

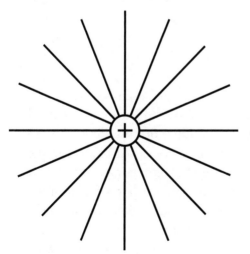

Notice that the lines are further apart the further away from the charge you go. Faraday used this geometrical fact to illustrate the physical fact that the force weakens the further from the charge you go

tive test particle would be attracted towards it, moving along these same lines of force but inward.)

Similarly, iron filings placed around a current-carrying wire suggested an underlying pattern of invisible lines of electrically induced magnetic force around the wire, and a moving magnet was assumed to create invisible lines of induced electric force.

According to Faraday, the lines of force around an electric or magnetic source represent the physical influence of the source, which permeates the space regardless of whether or not any test particle is there to demonstrate its effect. The lines therefore give a pictorial representation of the source's 'field' of influence. Most physicists paid no attention to what they saw as Faraday's naive idea, but of course Faraday was far from naive. He used his field lines to produce an alternative law to Coulomb's, a law with no possible connotations of action-at-a-distance: instead of relating the strength of the force between two electric particles, or between two magnetic poles, to the objects themselves – to the amount of charge or magnetism they possess, and to the distance between them, as required by Coulomb's law – he related it to the number of lines of force in the field around the particles. As you can see in the diagram of the lines of force around the electric charge, close to the charge where the force is strong, there are a lot of lines of force in any small area; similarly, there are a lot of lines of iron filings close to the poles of a magnet. In these regions of strong force, the 'density' of field lines – the number of lines passing through a small region of the field – is high. Further away from the fixed source, where the force is weaker, the lines are more spread out, so their 'density' is lower.

Although Faraday expressed this idea in words rather than

symbols, and his description of the field was qualitative rather than precisely quantitative, Maxwell realised that he really did have the fundamentals of a completely new, non-Newtonian way of analysing electric and magnetic phenomena. It was not yet ready to replace the Newtonian methods – Coulomb's law, for example, was a beautifully simple and accurate way of calculating static forces – but Maxwell took Faraday's foundation and turned it into a mathematical 'vector field'.

Geometrically, vectors can be represented by arrows, having both length and direction. In the case of the fixed electrically charged particle, for example, the electric force on the test particle, at any point in the 'field', can be represented by an arrow pointing in the direction of the force. The length of the arrow is proportional to the strength of the force at that point, as measured directly or as calculated from Coulomb's law.

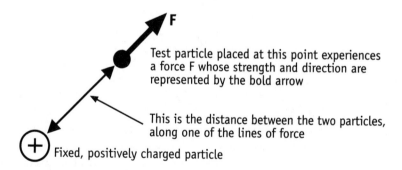

F

Test particle placed at this point experiences a force F whose strength and direction are represented by the bold arrow

This is the distance between the two particles, along one of the lines of force

Fixed, positively charged particle

The field idea supposes that the positive charge influences the entire space around itself, regardless of whether a test charge is in the vicinity; but the vector field is imagined by calculating the force that would act on a test charge if it *were* placed at various points in the area, and representing it by a vector. The whole *field* of force around the fixed positive charge looks like this (for selected points); the arrows get smaller the further away from the fixed charge, because the force weakens with distance.

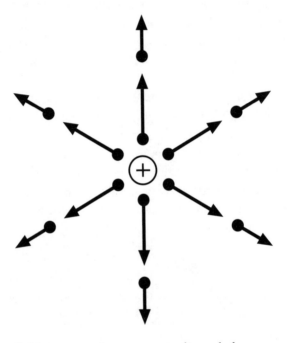

The field was much more complicated, however, when the central particle was no longer fixed, but was also moving, under the influence of the test charge or some other source; not only was the electric field changing as the particle moved through it, but there now existed an induced magnetic force, too, because of the particle's motion. To describe such an *electromagnetic* field, Maxwell needed to develop further his knowledge of vectors, and he sought help from one of the world's foremost authorities on the subject – his old friend Tait.

* *

As two-dimensional geometrical arrows, vectors had been used since the sixteenth century, but three-dimensional *algebraic* vector language was invented in the 1840s by the Irish mathematician, Sir William Rowan Hamilton. In Germany, Hermann Grassmann had independently invented a similar, *multi*-dimensional system at the same time, but he had used an unconventional symbolism which made his ideas difficult to grasp, and it took decades for them to be appreciated. Hamilton's vectors were part of a larger system called 'quaternions', but for simplicity's sake I will generally replace the word 'quaternion' with 'vector' in the historical passages below, because it was the vector part which Maxwell used.

Hamilton was a child prodigy who, by the age of ten, was familiar with eight languages, including Greek, Arabic and Sanskrit; at the age of 30, he was knighted, after a mathematical analysis he had made about the refraction of light from crystals was verified by experimental physicists. But he was a troubled soul. He had been orphaned as a child, and as an adult, he was prone to bouts of alcoholic overindulgence; like many pioneers, he suffered from feelings of doubt and isolation, although Tait's enthusiasm for vectors helped assuage these. During Tait's years in Belfast, he and Hamilton had established an important correspondence, in which they separately tackled problems involving the application of vectors to various physical situations. In April, 1859, after one such problem had been successfully tackled, Hamilton wrote joyfully to Tait, 'Could anything be simpler or more satisfactory? Do you not feel, as well as think, that we are on a right track, and shall be thanked hereafter? Never mind when ...'

In Belfast, Tait had applied vectors to some of Ampère's electromagnetic results, and this attempt inspired Maxwell in his search for the right tools to develop Faraday's ideas. The correspondence between Tait and Maxwell was vital to them both, according to Tait's laboratory assistant and biographer, C.G. Knott: 'We have Tait submitting his vector theorems to Maxwell's critical judgment, and Maxwell recognizing the power of the vector calculus as handled by Tait in getting at the heart of a physical problem.' Maxwell's response to Tait's use of vectors made a stark contrast to that of Thomson, who did not see what all the excitement was about. He fought vigorously (and good-naturedly) with Tait over the physical significance of vector language (particularly its quaternion form), he and Tait having begun co-authoring what was to become an important and successful textbook on mathematical physics; it was published in 1867, as *A Treatise on Natural Philosophy.*

Despite their mathematical disagreements, Tait and Thomson had become close friends and had great respect for each other. And for Maxwell, in whom they both recognised signs of greatness: Thomson later recollected a conversation they had had in Tait's small study (in which a shaded gas lamp on a cluttered table cast a shadow on walls already dimmed by tobacco smoke. 'It is when you are filling your pipe that you think your most briliant thoughts', Tait said). In this den of his, where he could contemplate and converse with scientific friends free of the demands of family or academic life, Tait would write in charcoal on the bare plaster walls, compiling a list of the greatest living scientists, in order of merit. Thomson recalled, 'Hamilton, Faraday, Andrews, Stokes and Joule headed the column if I remember right. Maxwell, then a rising star of the first magnitude in our eyes, was too young to appear on the list.' (He was in his early thirties at that time.)

Thomson and Tait's *Treatise on Natural Philosophy* was colloquially referred to as 'Thomson and Tait'; privately, they called it T and T' (pronounced T-and-T-dash), and the nicknames stuck: Thomson became T to his friends, while Tait was T'. Maxwell's nickname was dp/dt. This is a calculus term which appeared in an equation of thermodynamics that Thomson pioneered: dp/dt = JCM. The actual meaning of the complex thermodynamical terms 'dp/dt', 'J', 'C', and 'M' is not important here: the point was that 'JCM' are Maxwell's initials (James Clerk Maxwell), so his friends called him dp/dt in a mathematical allusion or in-joke. Using these shorthand names, the three Scottish friends dashed off letters to each other which give a marvellous glimpse into the process by which great science is done.

The letters between Tait and Maxwell are also full of dry and erudite humour, such as when Tait sent a 20-line rhyming poem asking Maxwell's opinion about an experiment he was planning. (Maxwell promptly replied with a 48-line rhyming poem giving detailed experimental advice!) Maxwell wrote to Tait about such flippant things as the 'danger of mutual destruction' due to the 'electricity of kissing', but more seriously (if no less humorously), while Tait referred to Maxwell as 'thou anti-distance-action sage', Maxwell called Tait 'the master of vectors':

> O T', I am desolated. I am like the Ninevites! Which is my right hand? Am I perverted? A mere man in a mirror, walking in a rain show? What saith the Master of Vectors?

This is a reference to a definition of vector *multiplication* – a definition that had obviously confused Maxwell. His letter highlights the fact that the basic grammatical rules of algebra, like multiplication, can be extended – albeit with difficulty, sometimes – to apply to other mathematical concepts than mere numbers. It is all a matter of coming up with appropriate new definitions.

It took Hamilton years to come up with the definition of vector multiplication, and when it finally hit him, like a bolt of lightning, he stopped right where he was, walking with his wife near the Brougham Bridge in Dublin, and carved his formula into the stone of the bridge. The secret lay in discarding the commutative law, xy = yx. This law holds true for all sorts of numbers, and mathematicians had assumed it to be a basic property of multiplication itself. Hamilton's genius lay in his realisation that you could still have multiplication without the commutative law, when you were multiplying other mathematical objects than numbers – objects such as geometrical arrows.

When you multiply two numbers, you increase one of the numbers a certain number of times to get a third number: 7 x 6 = 6 + 6 + 6 + 6 + 6 + 6 + 6 = 42. Similarly, you can multiply a geometrical arrow by a number, and increase its length in the same way: multiply a 6-centimetre-long arrow by the number 7 and you create a new arrow that is 42 centimetres long. But Hamilton's great discovery was a definition which allowed him to multiply one vector by *another vector*. That is a strange concept, and you might well ask, in company with Maxwell, what can it possibly mean to multiply an arrow by an arrow? The answer lies in the fact that an arrow has two characteristics, length and direction. To define the concept of vector multiplication, you would have to come up with some sort of grammatical operation that involved both these characteristics. Multiplying a vector by a number just changes the length of the arrow, but multiplying it by another vector, which most likely has a different direction from that of the first vector, must surely change the direction as well. (If the vectors are in the same direction, then their 'vector product' is defined to be zero.) The length of the new vector depends on the lengths of the two originals, but the new direction can be determined from a novel 'rule of thumb':

Call the two vectors **A** and **B** (bold face, upper-case letters are used here in order to distinguish vectors from arbitrary numbers or numerical values, which are represented by ordinary lower-case letters like 'x' and 'y', or 'a', 'b', 'c' and so on), and suppose they have the arbitrarily chosen directions shown.

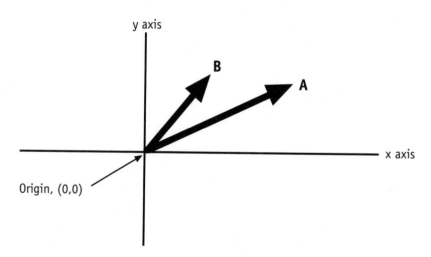

Assume the vectors are in the 'vertical plane': in other words, using your left hand, hold the page vertically in front of you. Then you can work out the direction of the product of these two vectors – the product being a new vector, **C**, where **A x B = C** – by stretching out your right arm, and bending your elbow as though you were about to arm wrestle. Your right forearm is vertical, parallel to the vertical 'y' axis, and your hand is open, fingers together, with your palm facing to your left. Now curl your fingers forward so that they are curling in the direction from **A** to **B**. Your thumb will point towards you, out of the plane of the page (and into the third dimension):

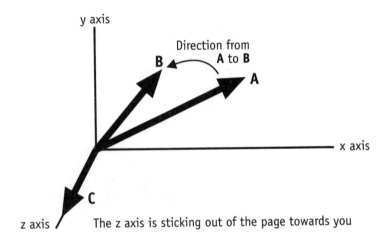

The z axis is sticking out of the page towards you

This is the rule Maxwell was referring to when he spoke of his confusion about his right hand; but the thing that seemed to unnerve him most about vector multiplication – and which made him talk about trying to distinguish his right hand from its reflection in the mirror – was the fact that it is not commutative, as you can see from the 'right-hand rule': to make your fingers curl in the direction from **B** to **A** in the above diagram, you have to twist your wrist around (anticlockwise) until your palm faces to the right, so you can curl your fingers from **B** to **A**; your thumb then points *away* from you rather than *towards* you (or into the page rather than out of it), so **B** x **A** points in the opposite direction to **A** x **B**. Opposite directions are distinguished algebraically by opposite signs (+ and –), so while **A** x **B** = **C** as shown, **B** x **A** = –**C**, not **C**.

In other words, a property of vector multiplication is that, while ordinary numbers multiply according to the commutative rule a x b = b x a, vectors multiply so that **A** x **B** = – **B** x **A**.

Thanks to analytic geometry, mathematicians are not limited to describing vectors by drawings of arrows or by creatively using their thumbs and fingers. Just as points can either be drawn on a graph or expressed algebraically, in terms of their coordinates, so can vectors be expressed both geometrically and algebraically. If you draw the arrow on a graph, with its end lined up with the origin, you can algebraically represent it by the point (a, b) that defines the tip of the arrow:

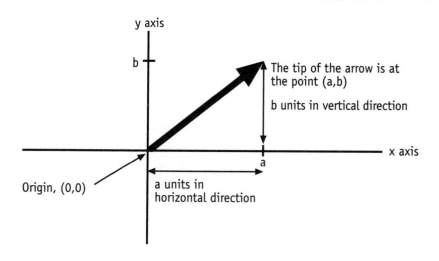

Two-dimensional vectors are visualised as being made from their horizontal and vertical 'components', 'a' and 'b'. Three-dimensional vectors have three components, 'a', 'b' and 'c', and are represented algebraically as (a, b, c); four-dimensional vectors have four components, and are represented as a quadruple of numbers (a, b, c, d), and so on, for as many dimensions as you like – a flexibility which makes vectors ideal for describing both Faraday's three-dimensional fields and Einstein's four-dimensional ones. Hamilton and Grassman showed that these arrays of numbers can be multiplied algebraically, according to specified rules relating their components. (In the three-dimensional case, the algebra gives both the magnitude and the direction of the new vector, but the right-hand rule is used when the focus is only on the direction.)

To create his ultimate field language, Maxwell needed to know about more complicated vector operations than multiplication – operations whose definitions were still so new and abstract that they did not yet have names. So he wrote to Tait, saying, 'What do you call this [operation]? I want to get a name or names for the result of it on vectors. Here are some rough hewn names. Will you, like a good Divinity, shape their ends properly, so as to make them stick?'

Being circumspect about the new language, he added,

> What I want is to ascertain from you if there are any better names for these things, or if these names are inconsistent with anything in vectors, for I am unlearned in vector idioms and may make solecisms. I want phrases of this kind to make statements in electromagnetism and I do not wish to expose either myself to the contempt of the initiated, or vectors to the scorn of the profane.

For one of the operations, which describes a rotational pattern in a field of vectors, he suggested the names 'twist' and 'twirl' (the latter he thought would be 'sufficiently racy'). Or better still, he suggested (because the operation is based on the definition of vector multiplication) a 'curl' (such as produced by the fingers in the right-hand rule). And for these other operations, he said, what about 'convergence' and 'slope'? Maxwell's terms did stick: 'curl' and 'divergence' (the negative of 'convergence') are now the standard English names for the vector operators he was enquiring about, and where he had suggested 'slope', mathematicians now use the equivalent term 'gradient' (or 'grad'). It is a wonderful revelation to read the thoughts in progress in Maxwell's private, informal letters. To witness the fact that mathematical definitions, and the understanding of their significance, have evolved like those in any human language, in a tentative, communal and often humorous process.

The curl, divergence and grad operators are defined algebraically – in terms of the algebraic definition of vectors; they also involve an extension of the language of calculus, which is another branch of analytic geometry. It was Maxwell's use of these vector operators that particularly impressed Einstein, who used an extension of vector calculus, called 'tensor calculus', in his general theory of relativity. At the time, though, Maxwell was a maverick in the world of mathematical physics.

But he knew that so far, the Newtonians – including some of the best mathematicians and theoretical physicists in the world, particularly Gauss, Riemann, Weber and Helmholtz in Germany, and Lagrange, Laplace, Ampère and Poisson in France – had not been able to provide a complete mathematical description of all the known electromagnetic phenomena; in other words, they had not yet done for electromagnetism what Newton had done for gravity. Maxwell believed their failure was because they had approached the subject from the wrong philosophical perspective – action-at-a-distance – and *with the wrong mathematical language*. The new language of (differential) vector calculus enabled him to break away once and for all from the limiting Newtonian view of electromagnetism, because it embodied the necessary paradigm shift. A shift of perspective that is startlingly simple, in hindsight.

A holistic paradigm

The fundamental calculus invented by Newton and Leibniz (and anticipated by others, including Archimedes) has two branches: 'integral calculus' (involving the operation of integration) and 'differential calculus' (involving differentiation). These two operations are inverses of each other, just as squares and square roots, or addition and subtraction, are inverses of each other: you can 'undo' the effect of an addition, for example, the addition of 6 to a given number, by a subtraction: $5 + 6 = 11$, but $11 - 6$ takes you back to the original number 5. Similarly, you can 'undo' the effect of an integration by applying a differentiation, and vice versa. Consequently, integration and differentiation are in a sense equivalent.

For instance, integration enables you to analyse mathematically a curved line by imagining it as being built out of tiny straight line segments; the smaller the segments, the closer the approximation to a single, smooth curve. Then you can do things like add up all the lengths of the line segments to calculate the length of the curve (the circumference of the circle in this case).

In other words, the concept of integration requires you

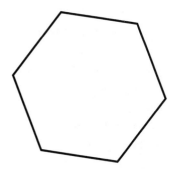
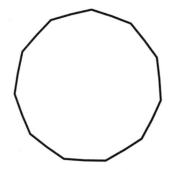

to conceptualise the curve in terms of tiny parts of itself, so that you start with the parts and end up with the whole. Differentiation works in the opposite way: you start with the whole curve, and then end up with the parts. Both methods provide equivalent descriptions of the same curve.

The Newtonian concept of a particle leads naturally to the choice of integration, because it initially focuses on the part – the particle itself. (On the single iron filing, for example, rather than the whole field.) Integration was the standard Newtonian approach to describing mathematically the effects of electromagnetism because it seemed entirely 'natural' to build the whole from the parts (to calculate the total magnetic force from the sum of its parts, for example). After all, that is how a machine or a dress or a cake is made. It took Maxwell to point out that – speaking philosophically rather than practically – the part is no more natural or fundamental than the whole, because both concepts require abstract thinking. For example, you can begin with points, and describe a line in terms of all the points it contains, or you can begin with lines, describing a point as the intersection of two such lines.

While the Newtonians focussed first on each force acting

in a given situation, and then used integration to calculate the combined effect of these forces, Faraday's initial focus was the whole (three-dimensional) field through which he assumed the individual forces were transmitted. Maxwell realised that differentiation was the appropriate mathematical technique for expressing Faraday's ideas – in particular, a special kind of three-dimensional differentiation called 'partial differentiation', which had been pioneered by George Green and Carl Gauss, among others. Partial differentiation was at the heart of the vector-calculus operators Maxwell had named. Using these operators, he was able to begin with the concept of the whole vector field, and then analyse it in terms of its parts, its vectors.

Maxwell's field vectors are (nowadays) called **E** and **B**; they essentially represent scaled versions of the measured electric and magnetic forces on 'test particles' at any point in the physical space around the electric and magnetic sources. In the earlier illustration of the forces around an electrically charged particle, the arrows were **E** vectors. In the left-hand diagram below, the arrows are some of the **B** vectors for the static magnetic case. The diagram shows how Maxwell used vectors to express Faraday's idea that these forces do not leap straight from the source to other objects, but affect the whole space around the source. The iron nail placed near the bar magnet is imagined to be affected by *the field of force* created by the magnet, not by a remote influence *directly* from the magnet itself.

By contrast, in the Newtonian representation, the space around the magnet is empty with regard to the action of the force, which acts between the magnet and the nail in the direction shown by the dotted line, but which leaps across that line without affecting the space it occupies.

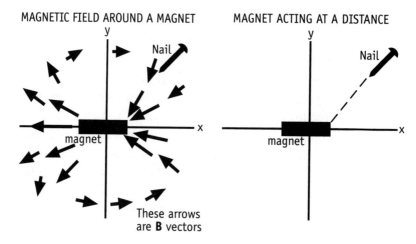

The field view assumes that forces cannot cross a space without affecting the intervening points of the space; it also assumes that electromagnetic effects do not occur instantaneously, but take time to get from one place to another. Maxwell expressed these assumptions in terms of the rate at which the electric and magnetic force vectors *change* as the force is transmitted. For example, notice that the **B** arrows change, in both size and direction, over the space between the nail and the bar magnet, starting off very large (and strong) at the pole of the magnet, and weakening over the distance to the nail. Electromagnetic forces may also change over time as well as space; for example, while a magnet is being physically moved through a loop of wire (in order to induce an electric current), the physical motion occurs in time as well as space, and so the magnetic field itself changes over time. (Imagine moving the bar magnet, and observing the iron filings realigning themselves as it moves: filings which were once close to the magnet are now further away, and vice versa, so the whole pattern of the field is continually changing while the magnet is moving.) Similarly, if an electric current changes in intensity, so that the electric charges are accelerating, then the electric field around the wire, and the induced magnetic field, also change in time.

Expressing rates of change over distance or time is one of the standard applications of differential calculus. With his newly named vector operators, Maxwell was able to express the rates of change of his vectors very elegantly; he thereby achieved not only a description of Faraday's field, but also something else that had long seemed impossible: a remarkably simple description of all the known experimental facts – that is, the discoveries of Oersted, Faraday and Coulomb. Oersted had discovered that an electric current creates a magnetic force as well as an electric one – a force that causes a magnet to *rotate*. (Faraday had used this rotation to invent

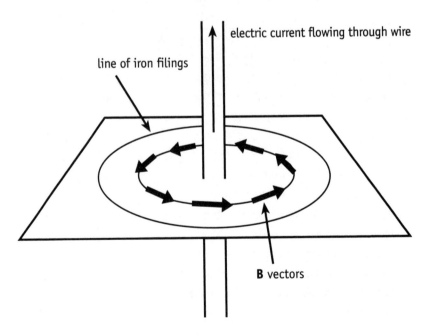

the prototypes of the electric motor and generator.) The idea can be illustrated by observing that iron filings scattered around a current-carrying wire align themselves in circles; the **B** vectors form a circular or 'rotational' pattern.

Maxwell visualised the magnetic field of **B** vectors as being 'wrapped around the current', and he expressed this image by saying the current produces a *curl* in the field of induced magnetic force vectors, **B** (this was the rotational effect for which he had coined the very name 'curl'). Similarly, Faraday's discovery, that moving a magnet through a loop of wire induces an electric field which causes a current to flow *around* the loop, could be expressed by saying a change over time in **B** produces a curl in the induced field of force vectors, **E**.

Maxwell then *generalised* Oersted's experimental result – something no one else had done – suggesting that an electric current was not the only source of magnetic induction: just as Faraday's moving magnet created a changing magnetic field which then induced an electric current, so should *any* changing field of electric force vectors, **E**, induce a magnetic field that 'curls'. (For instance, a discharging Leyden jar or a capacitor can create a 'movement of charge' analogous to that in an electric circuit.)

Finally, he used his convergence operator (which we now call 'divergence') to describe the two static cases – illustrated earlier in the diagrams of the **E** and **B** vectors around a fixed electric charge and a fixed magnet. This operator describes the tendency of the vectors to converge toward, or diverge away from, the central charge or magnet. In the diagram of the field around a positively charged particle, all the **E** vectors diverge away from the fixed charge, so the field as a whole has a 'divergence' which is found by applying the divergence operator. In the diagram of the **B** vectors around a bar magnet, however, the divergence operator says that the field's divergence is zero, because the vectors spread out from the left-hand side of the magnet and come back in at the right-hand side, giving a net 'spread' of zero.

And that is it. All that was known about electric, magnetic and electromagnetic forces could be expressed in four elegant equations: one involving the curl of **B**, one for the curl of **E**, one for the divergence of **B** and one for the divergence of **E**.

To derive these equations, Maxwell had used a much more complex analysis than I have implied here, and the meaning of the equations is also a little more complex than I have suggested. But the important and ironical point is that ultimately, by abandoning Newtonian mathematics – by starting with a whole vector field rather than with the individual electric and magnetic particles, and using differentiation rather than integration – Maxwell finally did for electromagnetism what Newton had done for gravity: he expressed it mathematically. His vector equations mathematically 'contain' and extend the Newtonian equations, including Coulomb's and Ampère's, and they link them into a conceptual framework that offers a philosophically satisfying explanation of *how* electric and magnetic forces are transmitted through space.

Maxwell's equations

At the time Maxwell wrote these equations, however, the Newtonians dominated mathematical physics, and he knew they would not be well disposed to his field theory of electromagnetism. The electromagnetic controversy was so protracted and intense that, as late as 1873, eight years after he had first published his radical theory, Maxwell spoke publicly and wryly of the field advocates being in the Newtonians' 'enemy's camp'.

In the *Treatise*, therefore, he tried to present his theory in conventional mathematical language where possible. He was particularly concerned not to alienate his readers with the technical symbolism of vector calculus, although he wanted the *idea* of a vector to be seen as fundamental in his theory. And he said that sometimes he used methods he thought were not the best in themselves, but which he judged to be more well known to his readers. (Newton, too, had been worried about being misunderstood by his peers, and in the *Principia*, he used traditional Greek geometry rather than his new calculus to derive his law of gravity.)

It also seems as though Maxwell bent over backwards to make clear every step in the development of his theory – to the point where he possibly included too much detail in his attempt to woo his colleagues. George FitzGerald, the great Irish mathematical physicist of the generation after Maxwell, called this detail 'the debris of his brilliant lines of assault, of his entrenched camps, of his battles'. Battles with the Newtonians, for whose benefit he included so much detail. (But FitzGerald had the luxury of hindsight – he wrote his comment after the field theory had been accepted, and the equations therefore needed no bolstering; Maxwell's careful detail could then be dismissed as 'debris'!)

It is true, though, that the four vector equations which contain the essence of electromagnetism are scattered over two chapters of the *Treatise* (and a separate paper), and are intermingled with a host of other equations used in the derivation of the theory. To modern eyes, even the four primary equations are expressed rather obscurely in the *Treatise*, in terms of either archaic vector symbols, or vector 'components'.

A simple example of a vector equation is $A = 2B + 3C$, where $A = (a_1, a_2)$, $B = (b_1, b_2)$, and $C = (c_1, c_2)$. In full, this equation is $(a_1, a_2) = 2 (b_1, b_2) + 3 (c_1, c_2)$, but it can also be written in 'component' form, in which this single ('whole') vector equation is replaced by two ordinary equations, one for the first components of each vector and one for the second: $a_1 = 2 b_1 + 3 c_1$, and $a_2 = 2 b_2 + 3 c_2$. Maxwell's two curl equations are much more complicated, and less transparent, than this, and they are three-dimensional, so in the *Treatise*, he represented one of them by three ordinary equations to make it more accessible to his colleagues. The other one he expressed in an early form of vector symbolism (although he expressed it more clearly in words, using his new term 'curl', in a separate paper). He also wrote his two divergence equations in component form.

It is partly because vector equations can be written in terms of their separate component equations that Thomson did not think whole-vector symbolism was important. In fact, when solving a vector equation, it is easiest to express it in component form, and Thomson did use vectors in this form. But, as Maxwell told Tait, the whole-vector symbolism gives an added conceptual insight that is missing when vectors are expressed merely in the form appropriate to calculation: 'The virtue of vectors lies not so much in *solving* hard questions, as in enabling us to see the *meaning* of the question and of its solution.'

This is because the symbolism is an integral part of the meaning of a mathematical equation. It has been said that mathematics enables you to make 'subtle conceptual deductions without having to think them out – all the thought has been built into the symbolism'. Vector calculus is an excellent example; the difference between it and its component form is somewhat analogous to the difference in being able to think in a foreign language rather than mentally translating everything back into your own language. In the latter case, you miss out on conceptual subtleties that may be untranslatable. In the case of vector language, being able to think in terms of the whole-vector symbolism means you can think in terms of the whole rather than the parts, and it is clear that the holistic paradigm governing the whole-vector calculus underlies the mathematical and philosophical development of Maxwell's final mathematical theory of electromagnetism.

Maxwell's most public expression of his belief in the conceptual power of vector language seems to have been a speech he gave on the occasion of Tait's being awarded a prestigious medal for two of his papers on vectors. Typically Maxwellian, it was a humorous as well as an insightful offering (the unexpected humour thoroughly mystifying the professor who had invited the comment, as Tait recorded with relish in his scrapbook, next to his copy of Maxwell's speech):

> The work of mathematicians is of two kinds; one is counting, the other is thinking. Now, these two operations help each other very much, but in a great many investigations, the counting is such long and such hard work, that the mathematician girds himself to it ... and thinks no more that day. Now Tait is the man to enable him to do it by thinking, a nobler though more expensive occupation, and in a way by which he will not make so many mistakes as if he had pages of equations to work out.

However, Maxwell acknowledged privately to Tait the need to remain 'bilingual' on vectors for a transition period, in order that the less elegant component form 'may help to introduce and explain the more perfect [vector form]', especially for the 'Cassios of the future' – those, like Shakespeare's 'great arithmetician' in *Othello*, who are more interested in calculating than conceptualising. Consequently, in his own *Treatise*, he chose to present mathematicians with 'pages of equations' rather than simply the four economical vector equations. When he did use vector symbolism, he used an early version of it (using Gothic letters for vectors rather than the modern bold-face Latin ones, and more cumbersome operational symbols).

Today, vector calculus is a standard and elementary part of mathematics, being introduced to students at junior undergraduate level. The modern development and application of vector symbolism is thanks largely to the work of Maxwell's brilliant disciple, the Englishman, Oliver Heaviside. He was an extraordinary, Dickensian character – a self-taught, working-class genius who was a telegrapher by day, until ill health and despair over class prejudice drove him into an early and impoverished 'retirement', in which he worked on mathematics and electromagnetic theory. He had learned much of his mathematics from 'T and T′ ', and he became one of the first to publicise the practical importance of Maxwell's mathematical field theory of electromagnetism, using it to solve some important telegraphic transmission problems.

A dozen years after Maxwell had written the *Treatise* (and twenty years after he had first published his theory), Heaviside singled out the four fundamental equations as being all that was necessary for Maxwell's theory, and expressed them in his new, simpler and more elegant vector symbolism. Similarly, nearly 150 years earlier, the Swiss mathematician, Leonhard Euler, had been the first to single out Newton's Second Law as the basic principle of mechanics. And Newton's successors replaced the symbolism he used in his calculus with Leibniz's more transparent notation.

While the symbolic form of Newton's and Maxwell's equations has changed over time, this does not mean that Newton and Maxwell are no longer the authors of these equations. The changes have made the mathematical content clearer, but they have not changed the mathematics itself. For example, the terms *curl* and *divergence* have the mathematical meaning Maxwell gave them, and are used in exactly the same way he used them, regardless of the choice of symbols in which they are expressed. However, there is no doubt that in terms of the modern symbols, Maxwell's equations are particularly beautiful:

$$\partial E / \partial t = c \nabla \times B - 4\pi J$$
$$\partial B / \partial t = -c \nabla \times E$$
$$\nabla . \, E = 4\pi \rho$$
$$\nabla . \, B = 0.$$

You can see just by looking at the first two (electromagnetic) equations that electricity and magnetism are intertwined: in each equation, **E** is on one side of the equals sign and **B** is on the other. The second pair of equations does not have this interrelationship, because it describes the separate electric and magnetic fields, including the cases of static electricity and magnetism. The symbols $\nabla . \, \mathbf{E}, \nabla . \, \mathbf{B}$ are the divergence terms, and $\nabla \times \mathbf{B}, \nabla \times \mathbf{E}$ are the curl **B** and curl **E** terms, respectively.

These equations are used by modern electrical engineers in the development of a huge range of electromagnetic technologies, yet they look like a four-line poem, the elegant pattern of symbols suggesting a kind of visual assonance through their intertwined partial repetitions. They call to mind another quatrain, from poet William Blake:

> To see a World in a Grain of Sand
> And a Heaven in a Wild Flower,
> Hold Infinity in the palm of your hand
> And Eternity in an hour.

Maxwell would use his equations to predict the existence of radio waves and the electromagnetic nature of light; through radiotelescopes, physicists can almost see infinity, and in Maxwell's combination of the single, arbitrary vectors **E** and **B** (the grains of sand) lies the blueprint for light and energy, and therefore for life itself.

MAXWELL'S RAINBOW

A bold mathematical prediction

Maxwell's use of differential vector calculus had brought Faraday's field idea to life, but the fact that it had enabled him to produce the first complete mathematical description of all that was known about electromagnetism did not, by itself, prove that fields were real and action-at-a-distance was not. The proof came because the equations predicted the existence of something that was then unknown, but which, when found to have a physical as well as a mathematical existence, proved once and for all that electromagnetism does *not* act instanteously, at-a-distance. This 'something' was the radio wave, which takes time to travel from a transmitter to a receiver. It was a phenomenon whose existence Maxwell teased out of the mathematical language itself.

By this I mean that, once he was satisfied with his description of the known facts, he began to analyse it to see what else it might contain. He could not have done this if his description had been in words rather than mathematical symbols. For example, if he had presented his account of Faraday's electromagnetic induction in words (as Faraday himself had done), he would have said something like, 'When there is an alteration (due to the motion of a magnet) in the number of lines of magnetic force which pass through a nearby loop of wire, a current-producing electric field is induced around the loop, which is measured by the rate of change of the magnetic force through the loop'. Linguistically speaking, there is nothing more contained in this expression of the experimental facts than these 40-odd words. You cannot apply the rules of grammar to create anything more than you put in. For instance, you cannot get a new slant on the situation by changing the tense of all the verbs in the sentence.

However, expressing the same facts in mathematical symbolism is not the end of it, because the rules of mathematics are more complex than the rules of ordinary grammar. As a very simple example, consider the equation $y = x$. It can be multiplied by 3 to produce a new equation, $3y = 3x$; this process does not actually yield anything new – it is the equivalent of converting all the verbs in the above sentence to the past tense. But this simple equation can also be turned into a differential equation ($dy/dx = 1$), by applying the more sophisticated grammatical rule of differentiation; the differential equation also contains the same information as the original, but in a different form, which emphasises a particular aspect of the information: dy/dx means 'the rate of change of 'y' with respect to 'x' ', so '$dy/dx = 1$' means that 'y' changes at the *same* rate as 'x' does – that is, the slope of this line is 1 in 1 (or 45 degrees). In other words, if you double 'x', then you are automatically doubling 'y', too. You can see this from the original equation, $y = x$, which tells you that when $x = 1$, then 'y', too, is 1 (because it is equal to 'x'), and when $x = 2$, then 'y' is also 2, and so on. But the differential form of the equation singles out for immediate attention the way 'y' *changes* as you move through space in the horizontal (or 'x') direction, whereas the original equation is general, and you have to *deduce* 'y' 's rate of change from it.

The deduction follows very simply in this example, because the original equation itself is so simple. But Maxwell's equations for the electric and magnetic field vectors are much more complicated than the equation of a single straight line because they represent a complex, sometimes changing pattern of generally *curved* field lines. I mentioned earlier that Maxwell's equations are already differential equations – recall that he used differential calculus to express the way the electromagnetic forces *change* throughout the field. But the equations can be differentiated *again*, and this time the extent of the 'rate of change' information they contain is not apparent, even to a mathematician, until the differentiation process has been carried out.

As it turns out, when Maxwell's equations are differentiated, something extraordinary happens: they become 'wave equations'. This mathematical fact would lead Maxwell to predict the existence of what we now call radio waves.

The one-dimensional form of the wave equation describes the waves produced when you pluck a guitar string. Or when you shake one end of a long, stretched, horizontal string or an uncoiled telephone cord and watch the wave run along it. By flicking the end of the string upwards, in the vertical direction, you 'disturb' it from its resting position, and it is this disturbance that is transmitted along the string as a wave.

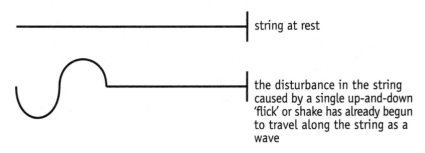

string at rest

the disturbance in the string caused by a single up-and-down 'flick' or shake has already begun to travel along the string as a wave

It is perhaps easier to visualise this particular kind of wave 'propagation' by thinking of ripples on a pond (although these are not quite the same kind of waves as those along the string). If you throw a stone into the water, it displaces the water around it, disturbing the pond's tranquil surface; the disturbance (and subsequent vibrations) propagate outwards as a series of waves or ripples. But the peculiar thing is that the disturbance is mainly in the up-and-down direction, although the ripples produced by it move in the horizontal (outward) direction. A leaf on the water will illustrate this process: the leaf is not carried outwards by the waves, but simply bobs up and down as each ripple or 'disturbance' passes by.

Similarly, the wave travels along the horizontal string, even though the disturbance is vertical. A wave like this, in which the directions of the wave's motion and the initial

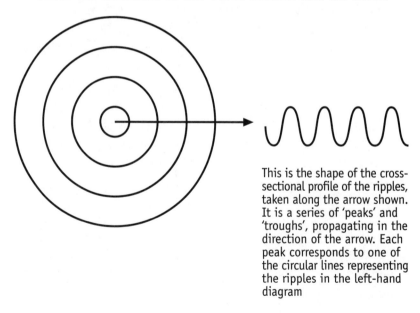

This is the shape of the cross-sectional profile of the ripples, taken along the arrow shown. It is a series of 'peaks' and 'troughs', propagating in the direction of the arrow. Each peak corresponds to one of the circular lines representing the ripples in the left-hand diagram

disturbance are perpendicular, is called a 'transverse' wave, and it is this type that the general 'wave equation' describes. (There are other kinds of waves, like sound waves, and their equations are variations on the theme of 'the wave equation'.)

Maxwell's unexpected equations were not referring to waves in water or string, but to waves in the electromagnetic field. Electromagnetic forces are created by moving electric or magnetic particles; one-dimensional forms of Maxwell's wave equations suggested that a change or 'disturbance' in such a force – due to a change in the particles' motion – is transmitted through the electromagnetic field like a wave along a string. In the 'plane polarised' case, in which only the intensity of the force changes (its direction remaining parallel to a given direction such as the 'x' axis), a 'snapshot' of the varying electric force vectors, frozen in time but still varying in space, would look like this:

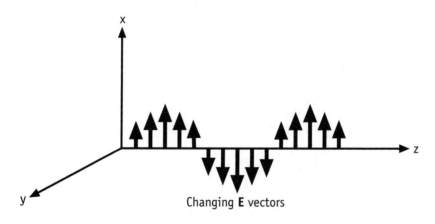

Changing **E** vectors

But that is not all. The changing electric force induces a changing magnetic force, in accordance with Maxwell's generalisation of Oersted's result. This changing magnetic force then induces a changing electric force, in accordance with

Faraday's law, and so on in an ongoing, 'leap-frogging' ripple effect. According to the full, three-dimensional version of Maxwell's wave equations, the wave of magnetic force is perpendicular to the electric wave, so that an *electromagnetic wave* has intertwined waves of successively induced electric and magnetic force.

It was a beautiful image, and it provided a beautifully simple illustration of Faraday's intuitive idea that electromagnetic effects do not act either instantaneously or at-a-distance: in this case, they are transmitted through space as a wave.

Maxwell had always had the feeling that the differences between the Newtonian and Faradayan perspectives could be resolved by the proper use of language. At last, he felt he had done it, and he was able to use his childhood bell-wire analogy to illustrate his achievement.

> When we ring a bell by means of a bellrope and wire, the successive parts of the rope and the wire are first tightened, and then moved, till at last the bell is rung at a distance by a process in which all the intermediate particles of the rope and wire have taken part one after the other.

In other words, it is not a *remote* force that produces an effect at a distance from a source, but one that has been transmitted point by point through the space. His original equations had expressed this idea in terms of changes in the field vectors, but his wave equation showed it more graphically: the intertwined, undulating electric and magnetic forces ripple from one point (or 'particle') in space to the next.

However, there was even more to Maxwell's wave equations than a neat resolution of the action-at-a-distance paradox. He realised, with growing excitement, that his electromagnetic waves were of the same type, *and had the same speed*, as light waves – a coincidence that suggested there may be some sort of physical relationship between these two apparently different phenomena.

At that time, light, too, was a mystery to physicists. While no one knew for certain anything about the deeper nature of electromagnetism, not even how it got from one place to another (instantaneously or otherwise), the only thing that was known about the fundamental nature of light was that it appeared to travel in waves. This was a conclusion based on the similarity of certain observed effects of light beams and water waves – in particular, the effect of 'interference'. When you throw two stones into a pond, the circular ripples spread out from each stone and 'interfere' or interact with each other to produce a more complicated pattern of 'peaks' and 'troughs'. Similarly, in certain circumstances, two interacting beams of light produce a pattern of bright and dark patches, which is identical to the crisscrossing pattern of peaks and troughs produced by the intersecting ripples. But no one knew what it was about light that was 'disturbed' or 'rippling'.

Until now. Emboldened by the implications of his equations, Maxwell took a radical leap: 'We have strong reason to conclude that light itself (including radiant heat, and *other radiation if any*) is electromagnetic.' The thing that is rippling in a light wave, he said, is the intensity of the electric and magnetic forces *which combine to create light itself.*

It was an awesome prediction. With the one set of equations, Maxwell believed he had revealed the secrets of two of the most intractable problems in physics: the fundamental nature of light (it is electromagnetic) and the resolution of the electromagnetic controversy (electric and magnetic forces do not act instantaneously at-a-distance, but act through a field, while changes in these forces are transmitted as waves that travel through the field at the speed of light). As Einstein would later put it, 'To few men in the world has such an experience been vouchsafed'. Maxwell himself, normally modest, had not been able to contain his excitement; he told his cousin he had a theory afloat which, 'till I am convinced of the contrary, I hold to be great guns'.

Unless electromagnetic waves could be physically detected, however, this beautiful theory would remain conjecture. And so it did for nearly 25 years. Then, in 1888, the German experimental physicist, Heinrich Hertz, found the first direct physical evidence of such waves. (In 1880, the American inventor, David Hughes, had claimed he had found such evidence, but the mainstream prejudice against Maxwell's theory was such that Hughes was dissuaded from pursuing his experiments.)

Maxwell's theory predicted that electromagnetic phenomena like light are composed of waves of electricity and magnetism, so it suggested that electricity must be able to travel not just along wires, like an electric current, but also through the air, through space itself, like light. Conversely, it suggested there might be other types of electromagnetic radiation than light, with different wavelengths from those of light. Hertz had studied Maxwell's theory, and also an attempt by his teacher, Hermann von Helmholtz, to remould Maxwell's equations into an action-at-a-distance theory. The discovery, or disproof, of the existence of physical electromagnetic waves would enable physicists to choose between Maxwell's theory and Helmholtz's version of it, because waves were impossible if action-at-a-distance were correct. (Recall that waves take time to travel – they are not transmitted instantaneously.)

Hertz's electromagnetic waves were produced by an oscillating electric spark. To physically generate such a spark, he used a Leyden jar (a battery produced a one-way current, rather than an oscillating discharge). As the spark jumped back and forth in a 'spark gap' made of two nearby metal balls connected to the jar by wires, its direction changed; according to Maxwell's theory, such changes or disturbances in the electric field should be propagated through the air as an electromagnetic wave.

The big moment came when, having set up, on the other side of the room, a 'receiver' made of a loop of wire with a spark gap, Hertz connected his 'transmitter' and generated a spark from the Leyden jar. Sure enough, another spark jumped back and forth in the spark gap of the 'receiver'. The electromagnetic radiation from the first circuit must have travelled across the room to the second loop and electrified it. The fact that it travelled as a wave was discerned when two sparks interacted with each other; the resulting interference patterns were characteristic of waves, with the wavelength of what we now call radio waves.

Hertz's electric waves also behaved like light in that they could be reflected, they had the same speed, and they cast 'shadows' when coming up against conducting metals while penetrating straight through nonconducting materials like wood or glass. Hertz was amazed; in his report of his discovery, he wrote that his waves could pass through a solid wall, and 'it is not without astonishment that one sees the sparks appear inside a closed room'.

In what Sir J.J. Thomson called 'one of the most marvellous triumphs of experimental skill and ingenuity … in the

whole history of Physics', Hertz had succeeded in demonstrating 'wireless' electromagnetic waves. Almost simultaneously, the English physicist, Oliver Lodge, exhibited
electromagnetic waves running along and around wires.
These discoveries provided the objective evidence needed to
validate both Maxwell's mathematics and Faraday's intuitive
field idea. They provided a dramatic illustration of the fact
that, despite what our senses tell us, electromagnetic effects
do *not* act instantaneously at-a-distance: waves propagate
from their source out into the surrounding space, and this
process takes time. Of course, using their physical senses
alone, physicists could never have discerned this transmission time, because a phenomenon travelling at the speed of
light travels some 300,000 kilometres every second. That is
why Maxwell's mathematics was so important: it showed
physicists where to look for the evidence they needed to
decide on the nature of electromagnetic transmission.

Ripples through history

Sadly, several years after his spectacular discovery, Hertz died
at just 37. Faraday, too, was dead. He had been plagued by
headaches and memory loss for decades, brought on, it is
said, by mercury poisoning during his bookbinding years,
and exacerbated by his habitual tendency to overwork, and
by the stress and disappointment of his struggle to get his
field idea accepted by the mathematicians. Nevertheless, he
had died, in 1867, at the relatively ripe old age of 75. He
had lived long enough to see Maxwell's field equations published in 1865, but he had missed out on Hertz's discovery,
and never knew that he, Faraday, would go down in history
as the creator of the now standard field idea. (And that his
would be the third portrait on Einstein's study wall.)

Neither Faraday nor Hertz lived to see the stunning practical use of Hertz's discovery made by the Irish–Italian physicist and entrepreneur, Guglielmo Marconi, who developed the first long-distance wireless telegraphy, sending Morse code not along wires but through the air, from one side of the Atlantic Ocean to the other. On 12 December 2001, the centenary of his achievement, widely called 'the birth of radio', was celebrated around the world. As with most great discoveries, however, there were a number of pioneers in the field: short-distance wireless electromagnetic transmissions (of varying type and success) had already been made by others, including the Americans Mahlon Loomis, Bradley Fiske and Nathan Stubblefield, the Serbian Nikola Tesla, the Russian Alexander Popov, the Britons Oliver Lodge and William Preece – and by Marconi himself. (There was also the problematic issue of how much Marconi had relied on other people's electromagnetic research and inventions, particularly Tesla's, to build his equipment – an issue which had also clouded Morse's achievements, and which not infrequently seems to arise during the process of patenting and developing commercially applicable ideas.)

Sound waves are not electromagnetic in nature – they are a series of physical compressions of the air or other medium transmitting the sound. They need to be electrically 'copied' if they are to be transmitted on electromagnetic radio waves. In 1892, Stubblefield had apparently succeeded in electrically transmitting sound over short distances, but on 24 December 1906, the Canadian-born engineer, Reginald Fessenden, broadcast the first long-distance sound-carrying waves, using his technique of 'amplitude modulation', which enabled electromagnetic 'carrier' waves to be modulated so they mimic sound waves. Such long-distance, sound-carrying electromagnetic waves are what is meant nowadays by the word 'radio', although the term 'radio waves' refers more generally to any electromagnetic radiation that has the frequency range used in radio broadcasts.

Today, radio waves are at the heart of modern communications technology, from radio, television and mobile phones to satellite dishes, and the huge radiotelescopes with which we communicate with the cosmos itself. But it all began with Maxwell's surprising wave equations.

The reason we can communicate on a cosmic scale – and the reason we can see starlight that has travelled billions of kilometres through space – is that, as the equations predicted, electromagnetic waves are self-perpetuating, because of the alternating processes of electromagnetic induction. These waves do lose intensity or energy in the radiation process, so that the light from distant stars is dim compared with that from the 'nearby' Sun, but they can still travel extraordinary distances. Similarly, light and other electromagnetic radiation emitted on Earth can travel into outer space: as you can see from the fact that light from a candle or light globe spreads out in a sphere, electromagnetic waves are like spherical versions of the circular ripples on a pond. Consequently, a radio or television signal beamed from a transmitter spreads out into our various living rooms, and also up into the sky, and then into space. Astronomers are currently analysing radio signals coming *in* from the cosmos, in the hope they might be from other intelligent species, somewhere out there in the vastness.

Physicists are also looking for evidence of *gravity* waves, such have been the extraordinary ramifications of Maxwell's field language. Both Faraday and Maxwell had realised that the field idea could be applied to gravity, although they did not pursue the application in detail. Einstein was the one who ultimately did it, in his general theory of relativity – whose description of four-dimensional spacetime had been motivated specifically by the search for a field theory of gravity.

He resolved gravity's problematic status as a force by replacing the 'magical', remote Newtonian conception with a continuous field of force, just as Faraday and Maxwell had done for electromagnetic forces. While the latter are imagined geometrically in terms of changes in the vector field around an object, gravitational forces are also pictured geometrically, in terms of the shape of spacetime itself – analogous to the curving of a trampoline when a person stands on it. The person's weight – the force of gravity acting on her – has affected the trampoline, distorting it from its flat shape, just as her presence is imagined to have distorted flat spacetime to create a static gravitational field around her. If she then jumps up and down on the trampoline, she will disturb its unmoving (static) surface, causing it to vibrate. Similarly, a vibrating gravitational source like a binary pulsar should transmit its 'disturbances' of spacetime as a series of gravitational vibrations or waves, just as a moving magnet or accelerating charge creates waves in the electromagnetic field. Physicists believe they now have indirect physical evidence of the existence of gravity waves, but the quest to measure them directly is a hot topic in current experimental physics.

Newton's equation of gravity is to Einstein's gravitational field equations what Coulomb's law is to Maxwell's equations – a static special case. However, in going beyond Newton's original vision of gravity, Einstein's theory is also even more accurate than Newton's. A major known limitation of the latter is that Mercury, the planet which experiences the strongest force of solar gravity because it is closest to the Sun, has an orbit that Newton's equation does not describe as accurately as it does the other planetary orbits. In weaker gravitational situations – those involving the Earth and its satellites, artificial and otherwise, and the planets apart from Mercury – Newton's and Einstein's equations are essentially equivalent, so Newton's is preferable because it is so simple. Closer to the Sun, or further out in the cosmos where there are massive sources of huge gravitational forces, like black holes or whole galaxies, Einstein's equations are needed.

As for electromagnetic waves, light and radio waves are not the only kind. Although he built them from only the then known facts about electricity and magnetism, plus a vital bit of imagination in generalising Oersted's result, Maxwell's equations turned out to describe a whole 'rainbow' of unexpected radiation, which he had referred to as 'other radiation, if any'. (By contrast, while an action-at-a-distance particle model was a useful way of visualising the effects of electromagnetic forces in many situations, it could never have predicted the existence of such radiation, which requires a description of how electromagnetism is *transmitted* through space – how it *radiates*.)

The visible part of the electromagnetic spectrum is made up of the various colours of light, each of which has a different range of vibrational frequency; that is, each colour emits a slightly different number of peaks and troughs in a given time period, with violet producing the most and red the least. This also means that red has the longest *wavelength* of the visible colours, and violet has the shortest, because the wavelength is the distance between two successive peaks.

Similarly, the frequency of vibration of all of Maxwell's electromagnetic waves determines their physical characteristics. At the high-frequency end are gamma rays, x-rays and ultra-violet light; in the middle are the colours of visible light; and at the low end are infra-red light, microwaves, and finally, radio waves.

Physicists now know that matter itself is electromagnetic. Atoms are made up of two different kinds of electric particle: positively charged protons and negatively charged electrons. Normally, the atoms in matter are electrically neutral – they have the same number of protons and electrons. Static electricity is produced by friction – by rubbing glass with silk, say, or rubbing amber with fur. It had been known since the work of the American scientist, Benjamin Franklin, in the eighteenth century, that these two processes produced opposite kinds of charge, in the sense that the properties of attraction and repulsion could be observed between such 'charged' objects. Before the twentieth century, however, no one knew what these 'electric charges' really were. Now we know that, while the atoms in normal matter are electrically neutral, when a substance like glass is rubbed with silk, some of the electrons in the atoms on the glass's surface are stripped off, so the glass is left with more protons than electrons, and therefore has a positive charge. (The extra electrons on the silk make *it* negative.) On the other hand, it is the fur that loses electrons when amber is rubbed, so the amber gains electrons from the fur and has an overall negative charge.

The individual electrons in atoms are not themselves static, but are constantly moving; this means they are *electromagnetic*: the fundamental sources of electromagnetism are the moving electrons in atoms, which induce magnetic forces, which induce electric forces, and so on. So magnetism itself is actually an electromagnetic phenomenon, induced by the moving electrons in atoms. (Ampère was the first to propose this idea, although he spoke of tiny electric currents within atoms, not of electrons.) Most substances are not magnetic, though, because the spinning motions of the electrons in their atoms cancel each other out – just as most substances are electrically neutral because the collective charges on the protons and electrons cancel each other out. However, just as rubbing a susceptible substance can give it an electric charge, so stroking a steel needle with a bar magnet causes the steel atoms to align themselves in such a way that the needle becomes magnetised. In practice, this process of magnetic alignment – that is, the process of magnetising suitable materials to create new magnets – is effected by an electrically generated (electro)magnetic field, rather than by the magnetic field of a bar magnet. Similarly, electricity (in the form of a current) is produced electromagnetically in a generator (or by the chemical release of the electrons in atoms, as in a battery) rather than by simple friction.

But Maxwell did not know about electrons when he built his theory of electromagnetism; despite the lack of knowledge at the time about the underlying nature of electricity and magnetism, he had been able to describe mathematically the known behaviour of these things in a way that is still relevant today. As for electromagnetic radiation, a phenomenon then completely unheard of, he predicted its existence purely mathematically. While Einstein's $E = mc^2$ would later show that matter and energy are intimately related, Maxwell's equations showed that electricity, magnetism, light, radio, x-rays, microwaves and other radiation are all interconnected. He had produced the first 'unified field theory', and ever since, physicists have been trying to find the ultimate one: the 'theory of everything'. Such was the power of Maxwell's use of vectors to imagine Faraday's fields.

A 'war' over vectors

Aside from its impact on the science of electromagnetism, Maxwell's *Treatise on Electricity and Magnetism* earned him a place in the history of mathematics as well as physics: in the modern classic, *A History of Mathematics*, Carl Boyer records the fact that Maxwell was 'influential in urging upon mathematicians and physicists the use of vectors' via his 'stunningly successful derivation of the electromagnetic wave equations'. He could not have done it without Tait, of course, who took Hamilton's ideas and polished them, and who was ever available to inspire Maxwell and answer his questions about the new language.

Thomson was not so impressed with the symbolic power of vectors (particularly in quaternion form), as I have already mentioned; late in life he wrote of his and Tait's collaboration, 'We have had a thirty-eight year war over quaternions'. A battle partly fuelled by the personality clashes between them. Thomson was driven, while Tait, although he was a hard worker, was a rather laid-back personality with a penchant for practical jokes. (Sometimes he would combine his sense of fun with his passion for golf by organising evening parties in which the game was played with phosphorescent golfballs.) Tait loved Maxwell's sense of humour, but he found Thomson too serious – charming though he was – and enjoyed making gentle fun of him. On the other hand, Thomson once wrote to Tait (about another mathematical topic) saying, 'It is not the thing I object to but your

Pecksniffian way of doing it'; according to Thomson, Pecksniff, a hero of Tait's, was a character of 'superhuman selfishness, cunning, and hypocrisy, splendidly depicted by Dickens'. (Tait also clashed with later vector *experts*, particularly Heaviside, because Tait was rather too rigid in keeping to Hamilton's original quaternion formulation. Thomson's disagreements with Tait were not all due to the *former's* stubborness!)

However, despite their frequent disagreements and their personality differences, Thomson and Tait remained lifelong friends; Thomson's blind spot on the full power of vectors was not merely due to this specific clash of personalities but was part of his general inability to 'absorb other people's ideas'. Perhaps he was always too busy: during his and Tait's collaboration on T and T′, he had been made a knight, an honour he received after his part in the successful laying of the first Atlantic telegraph cable. As a director of the board of the Atlantic Telegraph Company, he had been involved in the project financially, mathematically and practically.

He had been on board the ship when the first cable broke as it was being laid, and Maxwell had later written to him with a suggestion on how to let a cable down into the sea more gently, using a system of 'kites' tied to the cable to ease the tension on it as it was lowered from the ship. On the second attempt, Thomson was again on board as the ship steamed into a terrible storm that threatened to thwart the venture, but eventually the cable was laid, and Thomson sent the first signals along it. A month later, it broke, partly because of a problem in its construction, and partly because messages had been sent – against Thomson's mathematically based advice – at too high a voltage. There was a third failure, but finally, in 1866, the mission was successfully accomplished.

By 1885, over 100,000 kilometres of cable had been laid around the world; no wonder wireless telegraphy, or radio, would be such a breakthrough. As a professor at Glasgow, Thomson had trained a large proportion of the growing workforce in a profession that threatened to take over from religion as the most popular choice for graduating students: he noted there was 'quite an epidemic amongst the laboratory students of desire to become telegraphic engineers'.

Thomson became very wealthy from the combination of his scientific and business interests; he often worked on board his new, ocean-going yacht, on which he also entertained friends like Maxwell and Tait. In 1892, he was elevated to the peerage, as Lord Kelvin of Largs. Perhaps he was too successful to achieve the greatness for which he had so long seemed destined. Maxwell, on the other hand, was content with knowledge for its own sake. Like Faraday, whom Thomson had unsuccessfully tried to entice into a business partnership in which he could patent his discoveries and inventions, Maxwell gave his ideas freely to the world.

IMAGINING THE WORLD WITH THE LANGUAGE OF MATHEMATICS: A REVOLUTION IN PHYSICS

Maxwell's theory completely changed our perception of the physical and cultural landscape, so that nowadays it is hard for many of us even to imagine life without radio or television, or the efficient transmission of electricity. The theory's philosophical legacy was equally profound because it changed the way physicists perceived the relationship between language and reality, and thereby changed the way theoretical physics is done. As in most revolutions, however, the transformation came only after a struggle.

When Maxwell gave his electromagnetic theory to the world, few were ready for it. Not just because it opposed the Newtonian view, or because it used new vector language, but also because it was simply *too* mathematical. Physicists nowadays are used not only to describing the world but also to imagining it with the language of mathematics, but nineteenth-century physics was a more literal affair, and physicists just could not imagine a wave in a mathematical vector field. They needed a more concrete image if they were to believe in Maxwell's electromagnetic light waves. Even Hertz had to puzzle long and hard over the implications of Maxwell's theory, and really only came to understand it after he had seen for himself the physical, wireless waves it predicted. After all, theoretical physics had always been done by conceiving a physical theory in terms of concrete physical analogies, and then expressing those analogies mathematically so that quantitative predictions could be made to test the theory.

For example, in his theory of gravity, Newton treated the planets as though they were particles, with their gravity concentrated at their centres, and he imagined gravity to be a force, analogous to everyday push-pull forces; similarly, in attempting to create a theory of electromagnetism, physicists used various analogies to describe the behaviour of electric and magnetic objects. Having imagined these analogies, physicists then set about describing them, so that the role of the mathematical language in conventional theoretical physics was essentially descriptive, just as it is when describing actual experimental observations, or when inventing words to describe existing physical phenomena or ideas.

Maxwell himself used analogies – his first attempt at mathematically describing Faraday's 'lines of force' was based on a model of streamlines, like those in a river or other flowing fluid. But he had been thinking about the role of analogies in physics ever since his days with the Apostles at Cambridge, and in his breakthrough electromagnetic paper of 1865, and particularly in his great *Treatise*, he spelled out his position: physical analogies are useful for the initial development of a theory, as a 'temporary scaffold', but unless they have some basis in experimental fact, they have no place in the final theory itself. This is because there is too much danger of confusing the analogy with the reality, of believing the analogy *is* the reality. Of believing, as many did, that electricity actually *is* a fluid, rather than, say, a flow of tiny electric particles or 'electrons'; or that gravity is merely a push-pull kind of force, rather than, perhaps, a curvature of the geometry of four-dimensional spacetime.

On the other hand, the planets do look like particles, and because Newton had mathematically proved that the gravity of a spherical object does act from its centre – that is, from a single point – the particle analogy in that case is a good one. Similarly, it applies well to the observed motion of two nearby magnets, or two electric charges. The push-pull force analogy also works well at the everyday level of observation, and consequently, Newton's and Coulomb's 'particle-based' equations of force provide extremely accurate descriptions of the observed behaviour of everyday gravitational, electric and magnetic phenomena.

They do not say anything, however, about how such forces are transmitted from one particle to the other. All they say is that the *effect* of any one of these forces, *measured* when the particles are a given distance apart, is accurately described by the inverse-square law, which is based solely on properties of the particles themselves (their mass or charge or magnetic strength, and the distance between them). Unfortunately, most physicists had confused the mathematical equation with its apparent physical implication – that the force itself must act instantaneously, from a distance, if it can be measured and described in that way. But this was merely an inference, a physical analogy based on what *seemed* to be happening.

Faraday's field was also an analogy, based on the pattern of iron filings around a bar magnet or a current-carrying wire. But while the Newtonians tended to assert their action-at-a-distance analogy as physical fact, Maxwell presented the field concept only as a mathematical image, expressed in terms of *vector* fields; he did not presume to give a specific physical form to the field, by giving a concrete model of the invisible physical process by which it was supposed to actually transmit electromagnetic effects. He knew any such model would be mere speculation, because at that stage, there was no experimental evidence at all on how, or if, electromagnetism was transmitted. Maxwell wanted the mathematics to speak for itself, to guide physicists' imaginations without limiting them with a premature physical analogy.

However, when he presented an electromagnetic wave theory built out of the language of mathematics, most of his colleagues could not believe he was talking about reality at all. Thomson accused him of mysticism! And the great French mathematician, Henri Poincaré, said flippantly of Maxwell's *Treatise*, 'I understand everything in the book except what is meant by ... electricity' – which was a bit of an insult since the book was supposed to be a treatise on electricity. Poincaré, too, was complaining that Maxwell did not give enough physics in his theory. Hertz himself initially could not figure out what Maxwell was getting at, but eventually he understood, saying, 'What is Maxwell's theory? I cannot give any clearer or briefer answer than the following: Maxwell's theory is the system of Maxwell's equations'. (In other words, no physical analogy was needed to make it a 'proper' theory. Poincaré, too, eventually changed his mind, and was influential in introducing Maxwell's theory to his colleagues in France.)

But not even Hertz's discovery of physical electromagnetic waves convinced Thomson, who never significantly deviated from his original pronouncement, 'If I knew what the [mathematical] electromagnetic theory of light is, I might be able to think of it in relation to the fundamental princi-

ples of the [physical] wave theory of light'. He wanted
Maxwell's theory to be expressed in terms of a concrete
image rather than a mathematical equation: a mechanical
model like that which for decades had been used to illustrate
the original analogy between a light wave and a water ripple.
However, in putting his faith in this particular model,
Thomson was again being too impetuous.

A methodological paradigm shift: The slow transition from concrete models to mathematical imagination

It may seem surprising that physicists had believed in the
light wave analogy because, like gravity, electricity and magnetism, light also appears, to our physical senses, to travel
from its source to our eyes instantaneously. But it had been
known since Newton's day that light was not an action-at-
a-distance phenomenon because it is transmitted at a non-
instantaneous (or finite) speed. It was not Newton who
discovered and measured this speed, though, it was the
Dane, Ole Roemer. He showed that puzzling discrepancies
in the timing of the periodic eclipses of Jupiter's moons, as
observed on Earth, could be explained by assuming the light
from the moons took longer to reach the Earth when the
moons were further from the Earth than it did when they
were closer to it. This meant that light must have a finite
speed: if light propagated from the moons to the Earth
instantaneously, then the light received just before and just
after an eclipse should reach our telescopes immediately each
time, regardless of the changing distances between the
moons and the Earth as they travel in their orbits. Instead,
the light that signalled the timing of an eclipse was 'delayed'
when the Earth was further away from Jupiter, while it
arrived earlier than expected when the Earth was closer to

Jupiter. Knowing the time discrepancies, and the distances from Jupiter and its moons to the Earth, Roemer could estimate the speed – the distance per unit time – of light.

Newton had suggested that light was made up of particles, each travelling at this finite speed, and for 150 years, British physicists in particular had resisted, in deference to Newton, various wave notions of light. By the 1830s, however, they had finally accepted the wave theory proposed by the Englishman, Thomas Young, who, in 1801, had pointed out the similarity between the interference patterns created by light beams and by water waves.

Young had also been able to deduce the wavelength of light – a statistic to add to Roemer's speed – but even he did not know what light waves were actually made of. They were simply a physical analogy with water waves. Maxwell therefore urged caution in visualising them because they are 'founded only on a resemblance in <u>form</u>' (rather than substance). But for most physicists, the analogy had become fact, and they had set about imagining light waves in more detail – a difficult task because light can travel through a vacuum, through empty space, from Jupiter's moons to Earth, for example, or direct from the Sun itself.

This is different from other wave phenomena known at that time, like sound and water waves, which need a physical *medium* to transmit their effects. If you drop a stone into a pond, the disturbance it creates, due to the force of its impact, travels outward as ripples in the water. Similarly, sound travels from a source, such as your voice, by creating a series (or wave) of alternate compressions and decompressions of the air between your mouth and your listeners' ears, so it clearly fills up the whole space around you with its waves. In other words, just as the impact of the stone spreads across the pond, using the water as its medium, so sound needs a continuous, three-dimensional medium, like air or water, in order to get from its source to people's ears. It cannot travel through a vacuum. (I'm reminded of the sci-fi advertisement, 'Remember, in space, no one can hear you scream!')

The only way physicists could accept the growing experimental evidence in favour of light waves – evidence found by Young, the Frenchman Augustin Fresnel, and others – was to make *another* physical analogy. They suggested that, like water and sound waves, light waves need a physical medium through which to propagate. But because light *can* travel through a vacuum, through 'empty' space, the light medium must be physically undetectable – except, presumably, through its effect on light. Physicists called it the 'ether' (sometimes spelled 'aether'). It was a model based on a purely physical analogy with known wave phenomena, but since no ether had been physically observed, it was another example, like action-at-a-distance, of an interim analogy made on the basis of incomplete physical evidence.

Not surprisingly, then, it was a problematic idea to some physicists, just as action-at-a-distance had been to Newton and his contemporaries. But many were less cautious, especially those who accepted that electromagnetism also takes time to propagate, so that it, too, must need some sort of ether to help it travel. These physicisists – notably Henry, Faraday, Kirchhoff, Thomson and Gauss – had considered the possibility that electromagnetic effects propagate like light, even before Maxwell made his prediction; however, their speculations were informed hunches rather than a theory which offered a detailed mathematical description or blueprint of such a process. Some of them, including Faraday, had also wondered if there was some physical interconnection between light, electricity and magnetism. Faraday had discovered the first experimental evidence of such a connection in the 1840s: the 'magneto-optical effect' in which a magnet could change the orientation of light waves – a discovery which became the 'talk of the town' after it was announced at what has been dubbed the world's first science press conference.

But theoretical physics was so firmly rooted in the language of physical analogy that when Maxwell offered his unifying theory – which said that light is not only affected by electromagnetism, it *is* electromagnetism, that electromagnetic disturbances do not propagate *like* light, they *are* light, or are part of the same spectrum – most could not understand it. As far as they were concerned, any theory that purported to explain light needed to include a model of the mechanism by which it was transmitted – a model of the ether.

Maxwell himself had mathematically constructed an ingenious mechanical model of an ether, through which he imagined Faraday's field acted; it had a complicated system of imaginary vortices and gears which illustrated how observed electromagnetic effects, like the rotation of magnets by an induced magnetic force, could be mechanically produced and transmitted through the field. It was on the basis of this mathematical ether model that he first predicted the existence of electromagnetic waves, which happened to travel through the ether at the speed of light – something which surprised him greatly, according to a letter he wrote Thomson. He concluded from his calculations that light and electromagnetism propagated as waves in the same ether.

But Maxwell possessed extraordinary mental discipline;

he never lost sight of the boundary between theory and reality, metaphor and equation. He knew that, unlike Newton's fairly self-evident billiard ball model of planets moving around the Sun, his ether model, with its microscopic, mechanical flywheels and vortices, was purely imaginary, and might actually have nothing to do with electromagnetism. After all, while he could see a compass needle deflect in the presence of a current-carrying wire, and could obtain a measure of the *strength* of the induced force – he could not see the force itself; like all physicists of the time, he just had to guess at how it acted, at-a-distance, through a field, or otherwise. And he was not in the business of guessing.

His ultimate aim had always been to produce a theory whose physical assumptions had been fully verified, all other physical predictions arising purely out of the *structure of the mathematics which described the verified assumptions.* Since no one had shown experimentally that the ether even existed, let alone been able to observe how it worked, Maxwell did not want to include his hypothetical model of it as part of his final theory: if it were wrong, then any predictions arising from the theory might well be wrong, because they would have been made on the basis of the mathematical description of the incorrect model. Conversely, if electromagnetic waves *were* experimentally discovered, the ether model would be given an uncritical boost despite its preposterous complexity.

Maxwell's mechanical conception of the ether had helped him to get a handle on the problem of electromagnetic transmission, and to sort out the appropriate mathematical techniques for describing it, but he had only ever seen it as a 'temporary scaffold' for his thinking. So, with a philosophical rigour some say has never been bettered, he started again, focusing on the mathematics itself, rather than the physics.

It took him three more years, but he finally developed his vector language to the point I described earlier, where, to his enormous delight, his wave equations came out of the very grammar of his field equations – the equations with which he had described only the experimentally verified facts, not the imagined ether. In other words, in describing all that was known about electromagnetism, he also discovered the way it is transmitted, simply on the basis of the rules of mathematics. There was no need to guess at the complicated mechanism by which the ether was set vibrating as electromagnetic disturbances passed through it, because there was no real need of the ether at all: whereas ordinary physical waves *cannot* be described without reference to some medium of transmission, Maxwell's waves were described simply as waves in the intensity of the alternating, induced electric and magnetic forces as they spread through the field.

Consequently, no matter what its nature, the elusive ether was not necessary to an understanding of electromagnetic waves. Its only possible relevance to Maxwell's theory was as a reference point against which the speed of the waves could be measured. (The speed of an ocean wave, for instance, is measured with respect to the fixed seabed, and a fixed ether, pervading the whole universe, would provide the same kind of yardstick.)

While Maxwell had not assumed anything about the ether, he had not assumed any other connection between light and electromagnetism, either. And yet, the theoretical speed of his wave equations, calculated on the basis of purely electromagnetic measurements, turned out to be the same (to within experimental accuracy) as the known speed of light (measured with respect to the Earth). At last he was ready to propose his radical theory that light itself is an *electromagnetic* wave. A theory based not on a hunch or a spurious physical model, but on a logical chain of experimentally grounded mathematical reasoning. A theory expressed entirely in terms of a set of mathematical equations, which was the major reason the rest of the physics community had a problem with it. However, by presenting his mathematical equations rather than his physical model, Maxwell had implied that, in the hidden realms, mathematics was a more reliable guide to reality than physical intuition; that mathematical images like waves in a vector field might be as close as we can get to perceiving the more subtle aspects of physical reality. He had thereby created the methodology of modern mathematical physics: he said that while mathematics should keep its descriptive role in experimental physics, it should play a more creative role in theoretical physics. He called this new methodology a *mathesis* – a synthesis of experimental physics and mathematics, in which physical theories are created out of physical fact and mathematical grammar, rather than out of physical fact and physical speculation or unverified analogy.

Hertz concurred, saying he could not escape the feeling that mathematical formulas have 'an independent existence and intelligence of their own, that they are wiser ... even than their discoverers'. In Maxwell's case, his equations were wiser than he was in that he still believed in the 'probable' existence of a physical ether, even though his equations had made it all but superfluous.

But again Maxwell refused to leave it at a guess, and he came up with a clever idea which the German-born American physicist, Albert Michelson, developed into an ingenious method for physically detecting the ether's effect on light. Michelson was such a good experimentalist that he later won a Nobel Prize, but his superbly designed experiment, carried out in 1887 with fellow-American Edward Morley, did not detect the influence of the ether.

In 1905, Einstein turned this famous 'failure' into a success, in his new theory of relativity. He noted that Maxwell's equations, experimentally confirmed by Hertz and Lodge just after Michelson and Morley's experiment, contain the speed of light – denoted by 'c' in the equations. (It is the same 'c' that appears in $E = mc^2$; because the speed of light is so huge, the c^2 factor in Einstein's equation showed that a small amount of mass (m) could produce a huge amount of energy (E).) The ether hypothesis implied that light attains the particular speed 'c' only when it is measured with respect to its own medium, the ether, just as a swimmer's natural speed is measured with respect to the still water of a swimming pool. Einstein showed that this hypothesis had nonsensical consequences, and that the 'c' in Maxwell's (and his own) equations is a 'universal constant'. That is, it is the same for all of us, no matter how we are moving with respect to the light wave when we measure its speed. (Currently, it is an open question as to whether or not light has had the same speed at all times in the history of the universe, but on a human if not a cosmic time scale, 'c' is constant; it can be considered as a universal constant for all purposes except perhaps the study of the early universe.) This is an outline of the argument Einstein used to reject the idea of the ether:

The Earth is not fixed: it moves *through* space – and therefore through the ether – in its orbit around the Sun. This means that light signals sent in the same direction as the Earth is moving should pick up extra speed, in addition to their standard velocity with respect to the ether, just as a swimmer in a river picks up speed when swimming with the current. And just as the swimmer *loses* speed when swimming upstream, so light signals sent in the direction opposite to the Earth's motion should have a measured speed less than the ether speed 'c', while signals sent 'across-stream' would have a different speed again. (This was the gist of Maxwell's idea, and the Michelson–Morley experiment was designed to measure just such speed differences.)

In other words, experimenters would measure different speeds for light depending on which direction they were facing. Or on the time of day: because the Earth itself changes direction as it spins on its axis, if an experimental device sent a light signal from one end of the lab to the other, then over time this direction would change with respect to the direction of the Earth's orbital motion, ranging from 'downstream' to 'upstream' and everything in between, so the experimenter would measure different speeds of light depending on the time of day (see the following diagram).

This was not a satisfactory situation, philosophically speaking, but practically it had little relevance to daily life because the Earth's orbital speed is only about 30 kilometers per second, or 1/10,000 of the speed of light, so the downstream and across-stream speed differences due to the Earth's motion are tiny. (Which is why Michelson's equipment was so amazing – it was designed to be able to detect such tiny differences.) But Einstein had found a more serious consequence of the idea that the speed of light is relative.

For ten years, starting when he was still a teenager at school, Einstein had been puzzling over the problem of what would happen if he were riding along on a light wave. If he held a mirror out in front of him, both he and the mirror

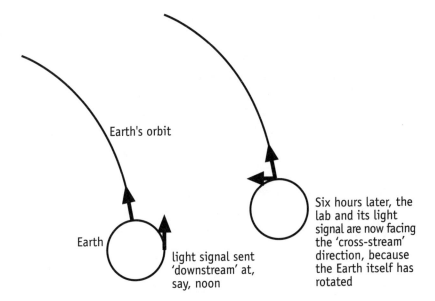

Earth's orbit

Earth

light signal sent
'downstream' at,
say, noon

Six hours later, the
lab and its light
signal are now facing
the 'cross-stream'
direction, because
the Earth itself has
rotated

would be travelling at the speed of light, and the reflected
light from his face would never quite be able to catch up to
the mirror, so he would not see his reflection in the mirror.
However, for us to make sense of the physical world, the
laws of physics cannot be relative. The laws of electromagnetism – such as those governing the ability of light to reflect
from your face to the mirror and back to your eyes – cannot
depend on how fast you are travelling.

To make the laws of light the same for everyone, the speed
of light had to be the same for everyone (as seemed to be the
case, according to laboratory experiments) so that even while
travelling on one light wave, Einstein would see other light
waves, reflected from his face, travel with the standard speed
of light (so they would catch up to the mirror in the usual
way). Similarly, a person on the ground would also see the
reflected light from Einstein's face travel at the usual speed.

Mechanically speaking, these are counter-intuitive ideas: you would expect the observer to see the reflected light leave Einstein's face at twice the normal speed, because it would have picked up speed from Einstein's own motion, just as the swimmer picks up speed from the river current. Similarly, if you are riding a bike with one arm outstretched, ready to catch a ball that someone standing on the road behind you is going to throw to you at the same speed with which the bike is moving, the ball will *not* catch up to your hand. So you would expect the reflected light from Einstein's face *not* to catch up to the mirror. Obviously, the laws of light propagation were very different from the Newtonian laws of ordinary motion, if Einstein's assumption that the laws of physics cannot be relative was correct. Could they possibly be integrated into a single theory, he wondered, or would the laws of electromagnetism always be separate from the laws of motion, as though there were two types of physics, one for ordinary matter and one for electromagnetism?

Maxwell had not been so concerned with these problems of relativity – his aim had been to describe the laws of electromagnetism as they had been experimentally observed by individual (stationary) scientists working alone in their labs. Now Einstein showed, in his theory of special relativity, that Maxwell's equations also applied in the bizarre world of super-fast relative motion, provided the speed of light in the equations were, indeed, a universal constant – the same for all observers – and not simply the speed relative to a fixed frame of reference like the ether. Which meant there was no need of the ether at all! (It also meant that it was the form of Newton's laws of motion, not Maxwell's equations, that had to be modified in relativistic situations, because relativity theory meant we had to abandon our everyday conception of time and space, no longer taking them for granted as the eternally fixed background of our physical existence: if the speed of light is always constant, then space and time must stretch and contract in certain situations to keep the equation balanced, because speed = distance/time.)

In effect, then, Einstein had argued against the need for a mechanical ether on the basis of the universal applicability of Maxwell's field equations alone, saying that a more dynamic view of space itself was all that was necessary to account for the motion of electromagnetic radiation through empty space. Not that Einstein's argument was immediately accepted; belief in the ether was so strong that experimenters, including Michelson himself, whose final experiment took place in 1929, kept replicating the Michelson–Morley experiment for another 40 years, each time 'failing' to find any significant variation in the speed of light. And mathematical physicists, including Thomson, kept creating elaborate theoretical models in order to fit the known electromagnetic facts into an ether framework.

Their dogged refusal to abandon the ether analogy showed just what an extraordinary thing Maxwell had done. He had not allowed his personal belief in the (probable) existence of the ether to prejudice his theory – he had let the mathematics speak for itself. Consequently, when the mounting experimental evidence – both against the ether hypothesis and in favour of Einstein's theory of relativity – suggested that a mechanical ether did not exist, Maxwell's theory remained unscathed.

In the end, then, despite his earlier praise for Faraday's simple, non-technical language, it was the ability of symbolic mathematical language to express the fundamental structures of nature 'without prejudice', as he put it, that Maxwell had used to such effect.

The Maxwellians

Einstein was one of the second generation of Maxwell's supporters, who helped demonstrate the soundness of his new methodology so that nowadays, physicists are completely comfortable with mathematical ideas like fields, waves of electric and magnetic intensity, and four-dimensional spacetime, and believe in them even though they may not be able to visualise them concretely. The *first* generation, including many of Maxwell's students, had worked hard to promote the electromagnetic theory itself, and to make it more accessible by refining and developing its physical ideas and consequences, and by expressing it in the simpler vector symbolism that had become more accepted since Maxwell published his equations.

They have been called 'the Maxwellians'. The most famous of them were the Germans Hertz and Helmholtz, the Austrian Ludwig Boltzmann, the Britsh contingent of Heaviside, FitzGerald, Lodge and J.J. Thomson, and the Dutchman Hendrik Lorentz. It was this group of mostly young physicists who – independently at first, before they heard of each other's work – consolidated and verified Maxwell's theory, teasing out all the physical implications of his equations (clearing away what FitzGerald called the 'debris') and applying them to the new electrical communications and power generation industries.

Working within a hostile establishment, the Maxwellians formed a grass-roots community of their own as they came together to support each other. Even the general public was hostile: when Lodge bought a copy of Maxwell's *Treatise*, the bookseller disdainfully muttered that it was 'a product of the over-educated' – which did not impress Lodge, who was currently studying part-time for an external degree, having left school at fourteen to work in his father's pottery business. Lodge had heard about the *Treatise* at the recent 1873 meeting of the British Association; the book had been praised in the presidental address, no less, but by a pure mathematician (Oxford professor H. Smith), not a physicist! It was an address which gave the eager young Lodge a mission: to make the 'production and demonstration of [Maxwell's] electromagnetic waves the work of my life'. Consequently, he immediately enrolled, at the relatively late age of 22, as a full-time student at University College to speed up his dream of becoming a physicist.

Five years later, he met FitzGerald in Dublin, and together they discussed plans for producing and detecting these waves. FitzGerald used Maxwell's theory to deduce a method of generating such waves with alternating sparks (the same method which Hertz later came up with, independently of FitzGerald). As Lodge put it, 'Henceforth it may be said that the remaining difficulty was not the production of the waves, but their detection', and indeed it was Hertz's painstaking demonstration of the wave nature of his radiation that so impressed J.J. Thomson and the other Maxwellians.

Because Lodge also managed to demonstrate electromagnetic waves at the same time as Hertz – although Lodge's waves were along wires rather than in free space – the British Maxwellians were immediately receptive to Hertz's discovery, while his Continental colleagues were not. But Lodge temporarily abandoned his own research in the wake of Hertz's success.

Lodge did go on to do more work on Maxwell's theory, including pioneering wireless telegraphy. But Heaviside was Maxwell's most fervent supporter. (Loner that he was, he, too, benefited from the community of Maxwellians, FitzGerald becoming a particular ally.) He had discovered Maxwell's work in the course of trying to solve telegraphic transmission problems; working as a telegrapher by day, he had taught himself enough mathematics to apply and extend Thomson's early theoretical work on the subject, but he soon realised that Maxwell's *Treatise* had more to offer. He was so impressed with the electromagnetic equations that he thought they would make Maxwell immortal, and that 'thousands of years hence', his soul 'will shine as one of the bright stars of the past ...'

One of his most important applications of Maxwell's theory was in the new telephone technology. Like the telegraph, the telephone was a marvellous application of electromagnetic induction. In essence, it worked like this: the mouthpiece contained a microphone, whose main component was a thin metal diaphragm; when you spoke into it, the sound waves produced by your voice impacted mechanically on the diaphragm, causing it to vibrate in synch. The moving diaphragm was attached to a magnet, and the now-moving magnet induced an electric current in a wire, the current oscillating in synch with the physical oscillations of the magnet – and also, therefore, with the physical sound of your voice. This 'analogue' current, whose wave profile was *analogous* to that of the sound waves from your voice, travelled down the connecting wire to the receiver of the phone at the other end. (Modern telephones still work in the same way, except that they generally use digital rather than analogue transmission.)

But since an electric current can also induce a magnetic force, this induced, oscillating current in turn induced an oscillating magnetic force, which alternately attracted and released a magnet attached to another metal diaphragm, hidden in the earpiece of the receiving phone. It was the exact reverse of the original set-up, so that the vibrations of the second diaphragm compressed and released the surrounding air in the earpiece, just as the original sound waves from your voice had done when you spoke into the mouthpiece. So the sound of your voice was recreated at the other end of the phone line – sound always being transmitted from its source via alternate compressions and decompressions of the air (or other medium). It was so simple, and yet so dramatic in its implications.

The Scottish-born American, Alexander Graham Bell, patented the telephone in 1876, sparking a legal challenge from simultaneous inventor and countryman, Elisha Gray. As with the invention of the telegraph beforehand and the radio still to come, litigation was a part of the early history of the telephone; Gray's was the first of many challenges by other contemporary inventors, including the Italian–American, Antonio Meucci. Meucci's suit lapsed when he died, but in 2002, the United States Congress passed a resolution acknowledging his claim – as it had done 60 years earlier on behalf of Tesla in his claim against Marconi. But the invention had not been an immediate success. Because of the symmetrical electromagnetic induction process, the early telephone used the one magnet and diaphragm set-up as both mouthpiece and earpiece, so that the user had to alternate listening and speaking. It was clumsy to use, and many people thought it was just a fad because it could carry a message for only a very short distance. In 1878, Maxwell gave a demonstration and lecture about this new-fangled invention, but the lecture theatre was too small to showcase the instrument because Maxwell's voice could be heard directly by his assistant without the need for the telephone! He turned the lecture into a very witty discourse which suggested that he did not see how the telephone could be of much use.

Fortunately, others had more faith. Bell, Edison and others made design improvements, including the inclusion of a battery-powered carbon microphone which enabled stronger signals to be sent to the receiver. The use of heavier copper wires helped improve the transmission of the electromagnetic signals between phones, and in 1887, Heaviside used Maxwell's own equations to help work out a better way of sending electric signals over longer distances, without too much distortion. Crucial to his analysis was Maxwell's idea of electricity existing not just in the wire itself, but also as electromagnetic waves in the field surrounding the wire. It enabled Heaviside to work out a way to minimise the dissipation of these waves as they travelled – an energy loss that resulted in a weakening and distortion of the signal. (Such an analysis was also vital in developing the efficient transmission of electricity from central power stations to homes and industries.)

But the engineers who controlled the industry would not deign to consider the word of an unknown, self-trained theorist, and refused even to publish Heaviside's paper – just as the previous generation of engineers had initially resisted Thomson's mathematical analysis of telegraphic transmission; in both cases, the theoreticians were right, and the prejudice against 'boffins' delayed the implementation of vital technological improvements.

The resistance to Heaviside's telephony paper was merely the latest in a long-running battle for the prickly outsider, who, as both a working-class telegrapher and a mathematical theorist, was not fully accepted by either group. Because of this, and because his health was poor – no doubt made worse by the stress of his struggle for acceptance, just as Faraday's crippling bouts of illness and memory loss had been at least partly due to his similar struggle – Heaviside had left his uncongenial job to work full-time on his research, unpaid apart from articles he wrote for the *Electrician* journal.

The rejection of his latest paper was a serious blow, but Heaviside would not be deterred; for him, science was a means of 'seeking the laws of God' and of 'banishing super-stition', and he saw himself as 'a mighty enthusiast, filled with a strong sense of my Duty to impart my knowledge to others and help them'. For such ends, this extraordinary man, living daily with poverty and rejection but never losing his passionate belief in the importance of his work, took the financial and intellectual risk of self-publishing some of his papers on electromagnetic waves (including one called *The General Solution of Maxwell's Electromagnetic Equations*), which he distributed to both British and Continental physicists. Fortunately, his timing was brilliant: it was the beginning of 1889, a year after his major telephony paper had been suppressed, but in the meantime, word had got around that Lodge and Hertz had discovered electromagnetic waves. The world was ready at last to pay attention to both Heaviside and Maxwell.

Maxwell had not been fated to meet any of his famous disciples or to see their verification of his theory; he had died, in 1879, of the same abdominal cancer that had killed his mother at the same age, just 48.

Three months before he died, he still showed his characteristic wry wit when he wrote about his pain in his last letter to Tait, whom he had nicknamed, with black humour, 'Headstone' (via the pun 'tête', the French word for 'head' which rhymes with Tait, and 'pierre', meaning 'Peter' or 'stone').

> While meditating, as is my wont on a Saturday afternoon, on the enjoyments and employments which might serve to occupy one or two of the aeonian aetherial phases of existence to which I am looking forward, I began to be painfully conscious of the essentially infinite variety of the sensations which can be elicited by the combined action of a finite number of nerves ...
>
> I have been so seedy that I could not read anything however profound without going to sleep over it. dp/dt

And then, the inevitable prognosis. Maxwell's doctor told him he had only a month to live. 'He took this with a calm and self-control which amazed the local physician ...' His specialist recalled that 'The calmness of his mind was never once disturbed. His sufferings were ... of a kind to try severely any ordinary patience and fortitude. But they were never spoken of by him in a complaining tone. In the midst of them his thoughts and considerations were rather for others than for himself'.

Maxwell told his doctor,

> I have been thinking how very gently I have always been dealt with. I have never had a violent shove in all my life. The only desire which I can have is like David to serve my own generation by the will of God, and then fall asleep.

Katherine and his friend and relative, Colin Mackenzie, were with him when he died. Maxwell asked Mackenzie to adjust his pillows and to

> 'Lay me down lower, for I am very low myself, and it suits me to lie low.' After this he breathed deeply and slowly and, with a long look at his wife, passed away.

After a memorial service at Cambridge, Maxwell was buried with his parents (and later, his wife) in a simple grave in a country church in Galloway. His lifelong friend, Lewis Campbell, had visited him at Glenlair not long before he died, and Maxwell had shown him, for the last time, his treasures from their shared childhood – including his paper on ovals – and had taken Campbell down to the River Urr, the scene of so many of their childhood idylls. It appears Tait did not visit his dying friend; he had become quite a recluse, leaving Edinburgh only to play golf at St Andrews, and besides, he hated emotionalism.

As for Thomson, Heaviside and FitzGerald clashed severely with him, because they felt he was hindering the progress of science by his refusal to accept Maxwell's theory. Thomson had always been a popular lecturer, but his class sizes began to dwindle after about 1875, despite the epidemic of interest in telegraphy, because students wanted to know about Maxwell's theory and Thomson did not teach it. He was becoming irrelevant to the younger generation, for all his lifelong fame and his phenomenal output of highly significant work. (In 1904, the young New Zealand-born discoverer of the nuclear structure of matter, Ernest Rutherford, wrote to his wife, 'Lord Kelvin has talked radium most of the day, and I admire his confidence in talking about a subject of which he has taken the trouble to learn so little'.)

Thomson had been the young Maxwell's mentor, but

clearly he did not take the trouble to understand his friend's mature work, thereby hampering its acceptance in the scientific community. It has been said that he never forgave Maxwell for rejecting the method of mechanical model-making on which he had based his life's work. Thomson's stubbornness brought him grief: towards the end of his life, he publicly and humbly admitted,

> One word characterizes the most strenuous efforts for the advancement of science that I have made perserveringly during fifty-five years; that word is FAILURE. I know no more of electric and magnetic force, or of the relation between ether, electricity and ponderable matter than I knew and tried to teach my students of natural philosophy fifty years ago in my first session as Professor.

Had Maxwell lived longer, who knows what else he would have accomplished. As it was, he achieved far more than his humble wish of serving his own generation, and his contribution to science has now been acknowledged with a memorial plaque in Westminster Abbey. Although he is not yet a household name, public interest in him has grown in the last decade or so, especially since 1993, when the James Clerk Maxwell Foundation bought and restored the house where he was born, at 14 India Street, Edinburgh.

For a hundred years, though, scientists have revered him. Max Planck, the great German theoretical physicist who created quantum theory at the dawn of the twentieth century, said Maxwell's theory 'must remain for all time one of the greatest triumphs of human intellectual endeavour'. Einstein said, 'One scientific epoch ended and another began with James Clerk Maxwell'.

But perhaps the most memorable tribute to Maxwell –

and to his belief that mathematics is the best language with which to imagine the world – is the t-shirt slogan seen on university campuses in the 1990s:

And God said,

$$\partial E/\partial t = c\nabla \times B - 4\pi J$$
$$\partial B/\partial t = -c\nabla \times E$$
$$\nabla \cdot E = 4\pi\rho$$
$$\nabla \cdot B = 0$$

And there was light.

EPILOGUE

Maxwell was born the same year in which his great predecessor Faraday had discovered electromagnetic induction, and he died the year his great successor Einstein was born. And just as Maxwell had based his theory on Faraday's ideas, so Einstein began with Maxwell's theory. In his revolutionary paper on special relativity, he said his aim was to give 'a simple and consistent theory of the electrodynamics of moving bodies based on Maxwell's theory for stationary bodies'. He found that Maxwell's equations correctly described the behaviour of light whether those observing it or the sources producing it were at rest *or* moving, a fact he used to reject the ether hypothesis.

But in an ironic twist, in the same year that he published his paper based on Maxwell's electromagnetic wave theory of light, Einstein published another paper in which he showed that light is not entirely a wave. In it, he consolidated Planck's mathematical hypothesis that on the subatomic level, light (and other electromagnetic radiation) sometimes behaves as if it were a particle – as if the subatomic structure of its waves were not smooth and continuous, but were broken up into separate portions or *quanta*.

In other words, at this level, electromagnetic radiation (such as light) exists as discrete particles or *packets* of energy. (This has been likened to an automatic bank teller, which allows you to take money out only in multiples of $10.) Planck had introduced the quantum idea as a mathematical technique for analysing a particular physical phenomenon (known as 'black body radiation'). Now Einstein was suggesting that quanta had a physical existence, not just a mathematical one; otherwise, certain observed phenomena cannot be explained – in particular, the photoelectric effect, in which certain solids give off electrons when irradiated with light. It was Einstein's analysis of this effect that led him to propose the existence of particles of light – later given the name 'photons' – which give up their energy to the electrons on a discrete, particle-for-particle exchange basis. A type of inverse photoelectric effect (called the photovoltaic effect) is used in modern solar cells for generating electricity directly from sunlight. The electrons given off when light falls on a silicon-coated metal plate produce an electric current.

As Maxwell realised, it is difficult and potentially misleading to try to explain, in terms of physical analogies and ordinary language, ideas that are essentially mathematical; but the concept that light is a wave which sometimes behaves like a particle is somewhat analogous to the idea that at the everyday level, a table top is smooth and solid, but if you could see the atoms of which its wood is made, you would see a lot of 'holes' – a lot of spaces between the electrons and other subatomic particles that make up the atom. Similarly, each wave or ripple on a pond radiates out from its centre by becoming larger and larger in a smooth, continuous way. One ripple follows another in a continuous train of waves – the distance between two successive ripples being the 'wavelength'. The quantum view would imagine each ripple to spread out in little bursts, or pulses, so that its motion looks smooth and continuous only when viewed from a 'distance'. No one doubts that the table top is solid and continuous, at the everyday level. And, as shown experimentally by Young and mathematically by Maxwell, on an everyday level, light is also continuous – it does radiate in a continuous wave. On the subatomic level, however, it appears in some situations to radiate in bursts or quanta.

In 1924, the young French duke, Louis de Broglie, gave a mathematical argument which suggested that all subatomic particles, not just photons, have a dual wave-particle nature. That electrons and other particles of *matter* (as opposed to radiation or *energy*) act as both waves and particles. In most everyday situations, electromagnetic radiation acts like a wave while matter is particulate; the radiation is described in terms of Maxwell's fields and the particles are described in terms of Newtonian mechanics. According to de Broglie, however, on a subatomic scale, radiation and matter were each considered to be *both* waves *and* particles. (Maxwell and Newton reconciled.) The German physicist, Max Born, and the Austrian, Erwin Schroedinger, developed de Broglie's idea, and now, along with photons of radiation, statistical, *mathematical* matter waves are at the heart of quantum physics. At the heart of physicists' understanding of the nature of electrons and the structure of atoms, and therefore of all the sciences and technologies which depend on an understanding of these things – particularly physics, chemistry, biology and microelectronics.

For example, the development of the microchip has resulted in the invention of household appliances, including computers, which have revolutionised the way we live. Quantum physics also plays a vital role in how we see ourselves biologically. It is because physicists have studied the structure of atoms simply for their own sake, simply in order to understand them, that biologists can now apply their results and their technologies (such as particle accelerators and computers) to the study of the atoms in the cells in our bodies, so they can build designer drugs to target specific cells, for instance, or study the structure of our genes, or the processes of damage to human cells due to illness or environmental factors like radiation. In fact, this kind of modern biological research is generally carried out by multidisciplinary teams of biologists, mathematicians, physicists and chemists.

Quantum theory has also led to the development of lasers, extremely concentrated beams of light which are used in surgery, compact disc players, laser printers, barcodes, industrial drills and scientific measuring instruments. Lasers can also carry telephone signals (along cables of glass fibre) more efficiently than ordinary electromagnetic signals can be sent along copper cables. The branch of quantum theory which deals specifically with electromagnetism, including light, is called quantum electrodynamics, or QED. It was created in 1927, when the English physicist, Paul Dirac, explicitly united, in the one unified field theory, quantum mechanics, special relativity theory and Maxwell's theory: Maxwell's equations for electromagnetic radiation also apply to the effects of a single photon – a *particle* of light. Dirac's equation for the behaviour of electrons has a similar mathematical structure to Maxwell's equations for photons, and together they form the foundation of QED.

This theory explains and describes the subatomic origins of electricity and magnetism, and their physical effects (such as the ability of atoms to absorb external energy and then emit it as light, as in lasers. The word 'laser' is an acronym for 'Light Amplification by Stimulated Emission of Radiation'). QED's predictions have been experimentally verified to such extraordinary accuracy that QED is considered to be the most successful physical theory ever. For instance, its mathematical prediction of the magnetic strength of a spinning electron has been experimentally confirmed to an accuracy of nine decimal places – that is, it differs from the measured value by about 0.0000000002. The famous American physicist Richard Feynman likened this degree of accuracy to mathematically calculating the distance between New York and Los Angeles to within a hair's breadth.

One of the more bizarre mathematical predictions of QED was the idea of 'antimatter' – specifically, an 'anti-electron', a particle exactly the same as an electron but with the opposite type of electric charge. At the time, the only positively charged particles known were protons, part of the nuclei of all atoms, but the proton is much heavier than the electron so Dirac thought he must have made a mistake. The German mathematician Herman Weyl took Dirac's mathematics seriously, however, and then in 1932, when 'positive electrons' or 'positrons' were physically discovered in 'cosmic rays' (high-energy radiation coming in from space), Dirac agreed that his 'equations had been smarter than he was'.

Rutherford (echoing Thomson's attitude to Maxwell's mathematical approach to physics) complained of Dirac's prediction, 'I would find it more to my liking if the theory had appeared *after* the experimental facts were established'. In other words, it was all right to use mathematics to explain known physical reality, but it was still seen widely as unscientific – supernatural – to predict mathematically the very existence of physical phenomena. But just as Maxwell's mathematical electromagnetic waves have led to practical applications like radiocommunications, so are Dirac's mathematical positrons manifesting themselves concretely;

for example, they are used as markers of atoms and cells in chemistry, biology and medicine, and in brain imaging for the detection of tumours or for understanding how the brain works. (When a positron meets an electron, they 'annihilate' each other, their matter being completely converted into photons, or light energy; the photons can be seen, thus marking the atom which produced the positron, thereby highlighting active or diseased areas of the brain.)

So Maxwell's nineteenth-century equations, a cornerstone of both QED and special relativity, survived the twentieth-century physics revolution, in which quantum theory and relativity completely changed our view of physical reality on both the microscopic and cosmic scales. Although additional equations are needed to describe quantum effects (and also, as Maxwell realised, the behaviour of ordinary charged particles as well as the fields they produce), Maxwell's equations apply accurately over a phenomenal scale, from the sub-atomic to the galactic. The microcosm united with the cosmos.

Newton's equations of motion and gravity, which apply with great accuracy at both everyday and astronomical levels, were the first to unify the universe in this way. And in quantum theory, the classical wave-particle controversy about the nature of light has been beautifully synthesised into an integrated whole, which also provides a solution to the old problem of how electromagnetic and gravitational forces get from their sources to other objects: on the microscopic scale, they are imagined to be transmitted from particle to particle by 'carrier' particles – photons and the still hypothetical 'gravitons' – while on the everyday and cosmic scales, they are transmitted through fields. Given the far-reaching power of this conceptual and philosophical unity, no wonder physicists are still hoping to find the unified 'theory of everything'. To combine quantum mechanics not only with electromagnetism and special relativity (as in QED), but also with general relativity.

At present, these two theories are not compatible. Quantum mechanics works on the minute, subatomic scale, and general relativity works on the cosmic scale, describing the origin and evolution of the universe itself; however, unlike Maxwell's equations, neither theory can be adapted to handle both scales. This means, for example, that although general relativity theory predicted the existence of black holes and the Big Bang, it cannot tell physicists what happens at the centre of a black hole, or what happened at the moment of the Big Bang. The whole theory breaks down at such tiny scales. For more than half a century, physicists have been trying to modify general relativity so that it can cope at the quantum level. If the speed of light is experimentally proven to have changed slightly over the past thirteen billion years, relativity theory will need a radical conceptual overhaul, at least insofar as it applies to the early universe – although its experimentally verified predictions will be preserved. Maybe someone will discover a new paradigm by which current physical theories can be transformed and made even more accurate, a paradigm which completely changes (again) the way we see physical reality.

Whatever happens, though, I am sure mathematics will be at the heart of it. It is a language that has enabled physicists to go beyond religion, philosophy, culture – beyond physical intuition itself – in their quest to understand the physical world we live in. It has been called a language of liberty, used by those whose fearless search for truth leads them beyond the conventions of their time.

Galileo and Newton believed mathematics, not philosophy or religious law, was the language of Nature; and in the process of establishing mathematics as the ultimate arbiter of physical reality, Maxwell, Einstein, Planck and their followers lived outside the conventions of the scientific establishment itself: it was well into the twentieth century before Maxwell's equations alone, unencumbered by attendant physical analogies and assumptions, were taken to be a sufficient description of electromagnetism (a reform initiated by Einstein, but not widespread until about 1930, more than 50 years after Maxwell had published his radical theory). It was then, too, that physicists began to accept the equations of quantum theory as fundamental, and stopped trying to understand it by 'agonizing appeals to a nonexistent physical intuition'.

Eventually, the paradigm shift was accomplished, and Dirac, reasserting Maxwell's mathesis, said that the sole purpose of theoretical physics is 'to calculate results that can be compared with experiment' – not to produce elaborate physical models in an effort to explain physical reality in ordinary words. It is a paradigm shift that has provoked some to wonder whether mathematics is somehow written into the very structure of the universe. One thing is certain, though. Mathematics is *our* language – a sublime creation of the human mind, built by countless generations from all parts of the world. A language of unity, both culturally and philosophically. It has given us a synthesis of waves and particles, energy and matter, language and reality, just as it embodies in its very grammar and symbols a synthesis of intellectual contributions from the East, the Middle East and the West. And it has changed forever the way we think about reality, and about our place in the universe.

APPENDIX

Proof that $\sqrt{2}$ is irrational

To prove that $\sqrt{2}$ is irrational, all you do is explore what would happen if it were rational — that is, if it could be written as $\sqrt{2} = x/y$, where 'x' and 'y' are positive whole numbers which give the simplest possible fractional version of $\sqrt{2}$. (For example, $^1/_2$ is simpler than $^2/_4$ although they are both equal to one-half.) The idea is to solve this equation, $\sqrt{2} = x/y$, and find appropriate values of x and y. An equation is 'solved' if actual numbers are found for 'x' and 'y', so that the left-hand side of the equation equals the right-hand side. But how can you do this if you do not know what number $\sqrt{2}$ is?

The trick is to square the equation — that is, you square the terms on each side of the equals sign: $(\sqrt{2})^2 = (x/y)^2$; that way, you are not changing the meaning of the equation because you have altered each symbol in exactly the same way, but you are rewriting it in a form which gives you 2 on the left-hand side of the 'equals' sign since by definition, $(\sqrt{2})^2 = 2$; you do not have to worry about what numerical value $\sqrt{2}$ actually has.

To summarise this procedure: square both sides of the equation $\sqrt{2} = x/y$ to get $(\sqrt{2})^2 = x^2/y^2$, or $2 = x^2/y^2$. You now want to see if you can find whole numbers 'x' and 'y' that 'solve' this equation. If you were doing this with a rational number, say $\sqrt{9}$, you would have: $\sqrt{9} = x/y$, which you would then square to give $9 = x^2/y^2$. If you let $x = 3$ and $y = 1$, the equation is solved: $9 = 3^2/1^2 = 9/1 = 9$ (so $\sqrt{9} = 3/1 = 3$).

Surprisingly, $2 = x^2/y^2$ is quite different. Rearrange it by taking the y^2 from the denominator (multiply both sides of the equation by y^2) so that the original equation $\sqrt{2} = x/y$ is now completely equivalent to the equation $2y^2 = x^2$. This means that x^2 is an even number, because it equals a multiple of 2 (that is, $2y^2$). But since 'x' is a positive, whole number, it must also be even, if its square is even. The even numbers, 2, 4, 6 ... all have even squares (4, 16, 36 ...), while the odd numbers, 1, 3, 5 ... have odd squares (1, 9, 25 ...). So to keep this clear in your mind, replace the 'x' in the equation $2y^2 = x^2$ by an obviously even number, 2p, where 'p' is another whole number (which is half of the original number 'x'). The equation now reads $2y^2 = (2p)^2$, or $2y^2 = 2p \times 2p = 4p^2$.

The equation $\sqrt{2} = x/y$ can now be written as $2y^2 = 4p^2$, and so the common factor 2 on both sides of the equation can be cancelled out to give $y^2 = 2p^2$. Now we have the same argument for 'y' as we had for 'x': it is a positive whole number, and since its square is even, it is even. Thus, the original 'x' and 'y' have to both be even if they are to satisfy the equation $\sqrt{2} = x/y$. But any rational fraction can be expressed in simplest form, without both numbers being even (or having other common factors). Therefore, $\sqrt{2}$ is not a rational fraction – it cannot be expressed as a ratio of two whole numbers. QED.

NOTES AND SOURCES

Any work of non-fiction is built on layers of knowledge which have been incorporated into the accepted knowledge of the discipline(s) in question, and I implicitly acknowledge all the thinkers and writers from countless generations who have built up and disseminated the subjects of mathematics, physics and philosophy.

With some influences, however, I can be more specific. In the following notes and sources, I have done my utmost to acknowledge not only all the direct quotations I have used, but also specific ideas or information that I am aware of having found useful in writing about the various topics in this book. The latter I include both in acknowledgment, and as suggestions for further reading about the topic at hand, because in the text I have had to simplify some of the concepts and topics discussed, or I have had space to present only snippets of information about an idea or a person, and some readers may want to explore these in a more detailed context.

In general, the references below give only the author's or editor's surname (or occasionally, only the title of the book), plus relevant page numbers; refer to the bibliography for full details of the source referenced. Anthologies are referenced by the author of the article in question, page numbers referring to the anthology.

A SEAMLESS INTERTWINING

* 'Could you lose it? ...' Malouf, (1993) p. 40. For Malouf's beautiful description of the differences in the way the settlers and Gemmy see the landscape, see pp. 67–8; see also pp. 129–30.
* Many thanks to Peter Biram for helping me in the process of learning to draw what I see.
* 'the world as we know it is in the last instance the words ...': Malouf (1985) p. 69. This quote was also used by Jopson, p. 5.

* $E = mc^2$, smoke detectors and EXIT signs: Bodanis, p. 193. Manufacturers are currently seeking non-radioactive sources for these objects.
* 'hidden and dimmer regions where Thought weds Fact': This quote is from Maxwell's paper 'On Bernoulli's Theory of Gases', see Jeans, p. 105; for Maxwell on mathematics and reality, see Harman (1998), pp. 4, 199.
* Einstein having Maxwell's photo on his study wall: Hoffmann, p. 46.

A RELUCTANT REVOLUTIONARY

* 'Mind of God': Hawking, p. 185; Davies, see book of this title.
* 'How can it be that mathematics ...': Einstein, *Geometry and Experience*, in *Ideas and Opinions*, p. 233.
* On the name 'Clerk Maxwell': It was effectively a hyphenated surname, and so you may see elsewhere references to Maxwell as Clerk Maxwell, and some older books actually list him alphabetically under 'C' rather than 'M'.
* **Key references on Maxwell:** A key scientific reference is Maxwell's own *Treatise on Electricity and Magnetism*. Unless otherwise stated, the references (below) to the *Treatise* are to volume 2, which contains Maxwell's theory of electromagnetism.

  For Maxwell's biographical details (in this and later chapters), I am particularly indebted to the following immensely readable and informative sources: Campbell and Garnett – referred to below as Campbell for simplicity; Tolstoy; Goldman; Maxwell's own letters as edited and annotated by Harman – these will be referenced below as 'Harman (ed.)'; and *James Clerk Maxwell: A Commemoration Volume*, which has a number of authors. I quote from the articles by J.J. Thomson, Einstein, Planck, Jeans and Lodge, who are listed separately in the bibliography.
* Tait's recollection about Maxwell 'reading old ballads, drawing curious diagrams ...' (Dafty): J.J. Thomson, p. 4.
* John's diary on James's ovals: Campbell, pp. 74–6. The other details about Maxwell's early life are also from Campbell.

BEETLES, STRINGS AND SEALING WAX

* Batteries may discharge even when not being used if there are impu-

rities in the zinc which set up a circuit in the cell in which tiny electric currents flow. See e.g. *The New Book of Knowledge*, p. 99.

* Lord Kelvin's house and electric light: Segre, p. 211.

Electromagnetism: A new and exciting science

* On Morse code: *The New Book of Knowledge*, Vol. 18 (T), pp. 51–2.

* On Morse and legal challenges to his patent (and a fascinating comparison of the motivations of the physicist versus the entrepreneur): Hochfelder. For a useful webpage on the history of telegraphy, see R. Victor Jones.

* On the large number of workers in the burgeoning telegraph research industry: McVeigh.

* On solar energy: Much current research is focusing on the development of solar *cells*, which generate electricity in a way analogous to that in a chemical battery, using light (rather than another chemical) to release electrons from molecules (of silicon); solar *panels* are used to collect heat for use in a generator.

* Maxwell's home laboratory, Harman (ed.) Vol I, and Tolstoy, p. 43

Newton's theory of gravity: Still producing exciting new discoveries

* On Airy's doubt over the exactness of Newton's law: Roy and Clarke, p. 100.

* Airy's praise of Maxwell's paper on Saturn's rings: Campbell, p. 249.

Leaving school

* Letters to Campbell: Harman (ed.), Vol. 1, pp. 68, 69, 96.

* Forbes's 'uncouth' letters: Harman (1998), p. 20.

THE NATURE OF PHYSICS

* Maxwell on physics: See his inaugural lecture as professor of natural philosophy at Aberdeen, Marischal College, 3 Nov 1856, Harman (ed.), Vol. 1, pp. 419–31.

* In talking about the fundamental grammar of the mathematical language, I mean the most basic, and the most widely *applied* part of the language – the algebraic part. But there are other branches of mathematics, such as set theory, which have different, essentially non-arithmetic kinds of rules.

* Einstein and the 'passion for comprehension': *Ideas and Opinions*, p. 342 (originally published in *Scientific American*, in 1950).
* Aristotle on physics: See extracts from *Physics (Book II)*, in Cahn, pp. 201–15. From his surviving works, it seems that Aristotle made more 'modern', experiment-based, scientific contributions to biology rather than physics.
* Oresme: Boyer pp. 263–7; Yang Hui: Mikami, p. 301.
* Gilbert as the first to systematically practice experimental physics: Tolstoy, p. 109.
* Newton to Hooke on 'shoulders of giants': Boyer, p. 392.

Experimental physics

* The unsung heroes of gravity include the sixth-century Greek philosopher, John Philoponus, who disagreed with the received Aristotelian wisdom that heavy objects fall faster than light ones. See Cushing, p. 74, and Boyer, p. 247.
* On the surface of the Earth, the acceleration produced by gravity is 32.1740 feet per second per second, or 9.8066 metres per second per second. These are often approximated by 32 feet per second per second, or, as I've done in the text, 9.8 metres per second per second.
* Crumpled paper experiment: Resnick and Halliday, p. 48.
* Apollo astronauts confirming Galileo's experiment: Davies and Gribbin, p. 38.

THE LANGUAGE OF PHYSICS

* For a history of the First Law, see Cushing, pp. 80–81.
* On $F = ma$: The numerical values of 'F', 'm' and 'a' are related in this way *for a given system of measurement units*.
* On $F = ma$ and F as an *effective* force: When talking about pushing furniture with a force 'F', 'F' is the effective force – your push minus friction from the floor surface. It is the same for gravity: when speaking of the 'force of gravity' pulling us to the ground, I really mean the force of gravity minus air resistance, as Galileo showed (although there is no upward air resistance against the downward pull of gravity unless you are actually falling). Only when the force of gravity acts alone is the associated acceleration constant. Gravity is such a powerful force, however, and air resistance is normally so tiny (unless you are talking

about something with a huge surface area in relation to its mass, like a feather or a sheet of paper) that air resistance can safely be ignored in most everyday calculations, because it does not influence the result of the calculation in any noticeable way. In the text, I will therefore generally ignore air resistance when I talk about the concept and effects of gravity.

* On Descartes' 'failure' to distinguish between weight and mass, see Maxwell's *Lecture on Faraday's Lines of Force*, in Harman (ed.), Vol. 2, p. 797.

* For more detail on the 'painful' evidence of the Third Law of Action and Reaction – and for all of Newton's laws – see Resnick and Halliday, p. 88. For Newton's quote, see Newton, *Principia* ..., Law III, in Fauvel and Gray, p. 390.

WHY NEWTON HELD THE WORLD IN THRALL

* On Newton's definition of gravity as a force: Nowadays, thanks to Faraday, Maxwell and Einstein, physicists usually call gravity a *field*, rather than a force, but in most everyday situations they still think of it as acting like a traditional push-pull type force – and therefore they still call it a 'fundamental force'.

* Accuracy of Newton's equations: Wigner, pp. 534–5.

* Aristarchus's heliocentric cosmology and ancient astronomers' objections to it: Boyer, p. 124.

* Historical overview of pre-Newtonian cosmology: For more information, see Rankin (on Newton and his influences); Sobel (in particular Galileo's trial and related circumstances, including Copernican heresy, p. 78, and the trial verdict, from which I have quoted, pp. 288–9); Wertheim (for general overview of physics history); Woolley (especially for sixteenth-century attitudes to science and mathematics).

 Note: On pp. 299–300, Sobel points out that there is a technical difference between the Church itself – including its leader, the pope – and administrative offices such as the Inquisition. Woolley, too, discusses the fact that it was not technically the Church who ordered the burning of heretics.

* Kepler used astronomical data collected by the famous Danish astronomer, Tycho Brahe.

How the universal theory of gravity was built: Imagining the world with the language of mathematics

* Dee on 'rays': Woolley, p. 53; Dee charged with 'calculating' etc: Woolley, p. 38; Dee casting horoscope for Elizabeth's coronation: Woolley, p. 59.

* 'There must be a drawing power': This is from an account of reminiscences by Newton towards the end of his life, by his young friend, William Stukeley, who wrote *Memoirs of Sir Isaac Newton's Life* (Taylor and Francis, London, 1936, pp. 19–20, quoted in Bodanis, p. 250); Stukeley's memoir was written in 1752, nearly a century after Newton had first pondered the falling apple (Resnick and Halliday, p. 382). For a popular account see also Rankin, Guillen, and Hatch.

* Diagram of satellite falling 'around' the Earth: See Rankin, p. 121, for an excellent related diagram.

* For more on Newton's final gravitational formula, see Hatch, p. 9; Resnick and Halliday, p. 383; Cushing, pp. 105–8.

Newton's theory of gravity in a nutshell

* Newton's powerful, single equation: When directions are taken into account, Newton's equation of gravity becomes a vector equation, which produces *several* component equations. See e.g. Fowles, p. 121f, for technicalities (including the solution of this equation to get elliptical orbits).

The difference between a theory and a hunch

* Hooke's notions of gravity and the inverse-square law: See the entry for Hooke in the *Dictionary of Scientific Biography*, pp. 485–6; Westfall, pp. 386–8; Hatch, p. 9.

* Hooke's analogy with the intensity of light: Guillen, p. 50, Rankin, pp. 50, 115.

Newton and his legacy

* "Man's reason should be made a judge over God's works". Maxwell's inaugural lecture at Aberdeen, Marischal College, 3 November 1856, Harman (ed.), Vol 1, pp. 419–31.

* Newton's biographical details: e.g. Westfall; Rankin; O'Connor and Robertson.

* Alternative explanation of Newton's letter to Hooke about 'the shoulders of giants', attributed to James Burke, in his PBS series 'Connections', in an online article 'Sir Isaac Newton', which I found at http://cns.pds.pvt.k12.ny.us/~jonathan/newton.html.
* Hobbes and the Plague: I owe this connection to Rankin, pp. 79–8; see also *Chambers Biographical Dictionary*. Hobbes's quote is from his book *The Leviathan* (see Cahn, pp. 449–507, for extracts).
* Voltaire on Newton's funeral: Boyer, p. 414.

RITES OF PASSAGE

* Maxwell's letter on hesitant speech: Campbell, p. 67.
* Physical description of Maxwell: Tolstoy, p. 53.
* Professor Swan's description of Maxwell: Goldman, pp. 54–5.
* For my portrait (here and below) of William Thomson, I am indebted primarily to Young and Wilson.
* Campbell's jealousy: Campbell, p. 151; 'genial and amusing': G. Tayler, quoted in Tolstoy, pp. 55–6. 'Maxwell as usual …': Campbell, p. 107
* 'I never met a man like him', and 'I would not have missed [walking with Maxwell] for anything': J.J. Thomson, p. 8.
* For the full version of Maxwell's poem, *Visions of a Wrangler* …, see Lancashire (ed.) or Campbell.
* Lawson letter: Campbell, pp. 174–6; Rev Butler ('lavish nature'): quoted in Tolstoy, p. 55; 'one of the best men': Campbell, p. 1.
* Berkeley's criticism of Newton's calculus: Boyer (1959), p. 225; (indeed, it took mathematicians another 200 years to make rigorous the difficult concept of 'infinitesimal' distances, which underlies calculus;) 'Infidel Mathematician': Kline, p. 145, Boyer, p. 430.
* Maxwell's philosophical influences: Goldman; Tolstoy; Harman (1998); Hutchinson (particularly regarding the influence of Darwinism, and Maurice).
* Laplace's 'Sire, I have no need …' quoted in Ekeland, p. 12, and Boyer, p. 494.
* Maxwell's Apostles essay on design in nature: Harman (ed.), Vol 1, pp. 227–8. Apostles essay on puns: Harman, Vol 1, p. 376; Apostles essay on analogies: quoted in Tolstoy, p. 77.
* On the history of the Tripos: Goldman, p. 17.

* For details on Tait, both here and below, I am indebted to Knott; e.g. 'I AM SENIOR WRANGLER', p. 9.
* The full version of Maxwell's poem, *Lines written under the conviction* ... is in Lancashire (ed.) or Campbell.
* Kirchoff on checking Maxwell's calculations: quoted in Goldman, p. 132.
* Maxwell's father's letter on Smiths Prize: Campbell, p. 207.

A FLEDGLING PHYSICIST
* Maxwell on his own carelessness: quoted in Goldman, p. 132.
* Maxwell's letters to Thomson: Tolstoy, p. 63, Harman (ed.), Vol 1, pp. 254–63, pp. 319–20.
* The dog's eye: Harman (ed.), Vol 1, p. 308; Toby at the Cavendish: Campbell, p. 369; Maxwell on killing animals: Campbell, p. 155.
* Maxwell's letters to Munro: Campbell p. 210.
* Thomson's first electric paper: Cushing, p. 188.
* Maxwell to his father on 'poaching' on Thomson's electrical preserves: Harman (ed.), Vol 1, p. 325.
* Maxwell and the working men's classes: Harman (ed.), Vol 1, letters 82 and 98; Campbell, p. 291.
* Maxwell's letter on novels: Harman (ed.), Vol 1, p. 328; on Charlotte Bronte: Campbell, p. 190.
* Maxwell's letter to Litchfield (on Pomeroy's death): Tolstoy, p. 87; Tayler's recollection of father and son: Campbell, p. 174.
* The full version of Maxwell's poem on his father's death, *Recollections of Dreamland*, is in Campbell.

ELECTROMAGNETIC CONTROVERSY
* Maxwell's letters to Campbell and Thomson on his father's death: Harman (ed.), Vol 1, pp. 405–6.
* 'No jokes here': quoted in Tolstoy, p. 81, (from Campbell).
* The full version of Maxwell's poem to his wife is in Campbell.
* Crimean War: My account of the role of Russell and Nightingale is based on that in the *Oxford Reference Encyclopedia*.

History of the controversy
* Priority dispute between Newton and Leibniz, and their followers: For

a brief, extremely unflattering summary of Newton's role in the dispute, see Hawking, p. 192. For more details, see Hellman.

* Leibniz on God, empty space and monads: Leibniz, pp. 216, 218; Newton reducing God to a watchmaker, *ibid.*, pp. 216–17.

* Descartes on vortices: Rankin, p. 74; for an idea of Descartes's philosophy see Descartes (in Cahn), e.g. proof of the existence of God, p. 428.

* Voltaire satirising Leibniz's optimism: *Candide*, p. 57.

* My version of Newton's quote on the absurdity of action-at-a-distance is taken from a lecture given by Maxwell in 1873, *Lecture on Faraday's lines of force*, Harman (ed.), Vol. 2, p. 798; Faraday's knowledge of this quote: Agassi.

* Maxwell's acknowledgment of Thomson is from the preface to the *Treatise* (Vol. 1); see also Cushing, pp. 189–90, Segre, p. 133; Goldman, p. 137; Maxwell on Faraday's mathematical language: *Treatise*, Vol. 1 (quoted in Agassi, p. 306), and *Treatise*, Vol. 2, p. 176, article 528.

* Airy's dismissal of Faraday's lines of force: quoted by Resnick and Halliday, p. 852, and J.J. Thomson, p. 28.

* Tait asking Maxwell about his new electromagnetic paper, Harman (ed.), Vol. 1, p. 558.

* Faraday's letter to Maxwell on 'On Faraday's Lines of Force': Goldman, p. 143 from Campbell.

A working-class dreamer

* Faraday's disillusionment with the scientific establishment, and related quotes: Agassi, e.g. pp. xi, 77–8, 135–52, 191.

* On Sandemanianism (including the Bible's 'plain style'): Cantor, p. 73.

* Official account of Faraday's early struggle (including rag shop, Tatum's lectures and Faraday's English and elocution lessons): Memorandum to Robert Peel, in James (ed.), pp. 244–6.

* Humphry Davy's origins: Knight, p. 43; Bonaparte prize: Guillen, p. 141. See also *Dictionary of Scientific Biography*.

* 'As many coals and candles as he wanted', memorandum to Robert Peel, James (ed.), p. 246.

* Faraday's personality: Agassi, p. 331.

* Cornelia Crosse and Juliet Pollock on Faraday's lecturing style, and the quote from commentator (Tyndall) on Faraday's 'magic': Forgan, pp. 62–3.

Faraday and mathematics

* Faraday on his lack of mathematics: Williams (ed.), pp. 7, 132, 134, 138, 154 and re geometry, p. 295; on occasionally being glad not to be a mathematician, p. 292.
* Faraday's letter to Maxwell asking him to express his ideas in 'common language': Campbell, pp. 288–90; Maxwell's response: letter to William Thomson, Harman (ed.), Vol. 1, p. 556.

Faraday's fields versus Newtonian particles

* On Faraday's lines of force: For a more detailed account of Faraday's ideas (including their debt to an earlier use of iron filings by William Wollaston) see Agassi, e.g. pp. 91–4. For Faraday on the transmission of forces from particle to particle, see Agassi, p. 262, quoting Faraday's *Experimental Researches*, paragraph 1166.
* Voltaire's 'furious contradictions': *Lettres Philosophiques* (Fourteenth Letter), pp. 90–4.
* Thomson's 'singularly winning smile': Wilson; J.J. Thomson on his being 'a good radiator but a bad absorber': Young, p. 16.
* Thomson on the 'race' between himself and Maxwell: Harman (ed.), Vol. 1, Introduction, p. 24.

MATHEMATICS AS LANGUAGE

Mathematics and ordinary language

* 'Queen of science', 'servant of physics': These words come from the title of Eric Temple Bell's classic history of mathematics, *Mathematics: Queen and Servant of Science*. Gauss had used the term 'queen of science' to describe the growing importance of mathematics in physics at the beginning of the nineteenth century, but it would become even more appropriate with Maxwell's work, and that of subsequent theoretical physicists.
* Definition of force quoted is from the *Oxford English Reference Dictionary*.
* Rainbows and hanging chains: Thompson, p. 269.
* The idea of nuclear fusion and its relationship to the Sun followed from the pioneering work (in the US in the 1920s) of people like Cecilia Payne, Henry Norris Russell and Robert Atkinson.
* Einstein on Hiroshima (becoming a shoemaker): quoted in Schwartz

and McGuinness, p. 166. *How Is It Done?* translates this quote using 'watchmaker' instead of 'shoemaker', p. 177.

A language of pattern

* Widman and '+' and '–' signs (and Descartes and the popularisation of '+' and '–' symbols): Boyer, pp. 280, 337.

* The 'unreasonable effectiveness' of mathematics: Wigner, pp. 534–5.

Pure mathematical language

* On mathematical self-consistency and proof: This is true, as far as most mathematics is concerned, but in 1931, the Austrian mathematician, Kurt Godel, showed that theoretically, not all mathematical propositions can be proved or disproved using the usual methods. See Boyer, p. 612, for brief discussion of attempts to get around the implications of Godel's theorem from *outside* the arithmetic basis of mathematics.

* Any positive number has *two* square roots (e.g. both 2 and -2 are square roots of 4), because the product of two negative numbers is positive (as it is linguistically in 'I will not not go').

Geometry: Pictures and proofs

* Theano and Perictione quotes: Meurier; Greek law forbidding women's attendance at public meetings: Cameron, p. 1.

* Newton on the Pythagoreans and cosmic harmony: Sworder, pp. 70–1; Wertheim, pp. 124–5.

* Euclid's *Elements* 'fabulously successful', and over a thousand editions: Boyer, pp. 100, 119.

* Proof of angles in triangle: Euclid said the angles in a triangle added up to 'two right angles', Boyer, p. 107; the proof I have given uses the later notion of degrees, see e.g. the Year 10 textbook by Lynch and Parr, pp. 198–9.

Calculating the size of the Earth

* Eratosthenes's calculation of the circumference of the Earth, and Aristarchus on the distance and size of the Moon, Boyer, pp. 159–61, and, in more detail, Resnikoff and Wells, pp. 93–103.

* <u>Diagrams:</u> I have adapted my diagrams for the scaphe, and for the Sun's rays as they hit the Earth and its shadow sticks, from those on p. 94 of Resnikoff and Wells.

* Regarding modern measurements, and the effect of the flattening at the poles on the geometry, see *How Is It Done?* p. 367; the use of degrees not then being widespread: Boyer, p. 159.

* Using a coin to estimate the distance to the Moon: e.g. Vorderman, p. 96.

Geometry and algebra unified

* 'The cold breath of Rome': Boyer, p. 175; 'the spirit of mathematics languished …': Boyer, p. 194. I am indebted to Boyer for my overview of the history of mathematics, see especially pp. 175–7, 187–9.

* Re Pythagoras's theorem and algebraic geometry: Khayyam and his colleagues treated much more sophisticated problems than this – I have only used it as an accessible illustration of the idea. However, the theorem has actual historical relevance to the work of the ancient Babylonians and Chinese, who treated it purely algebraically. But they did not have a sophisticated form of geometry, like the Greeks, so their algebraic treatments of this theorem do not imply a conscious synthesis of algebra and geometry, such as the later Arabic and European mathematicians produced.

* For more details on Indian, Chinese and Arabic mathematics and mathematicians, see Boyer, pp. 195–245.

THE MAGICAL SYNTHESIS OF ALGEBRA AND GEOMETRY

* On the Greeks' view of mathematical coordinates: Boyer, p. 156.

Four-dimensional geometry

* Minkowski calling Einstein a 'lazy dog': Hoffmann, p. 85; Minkowski's geometrical reformulation of special relativity: Schutz, p. 1.

* Flat and curved spacetime – special and general relativity: The geometrical difference between the two theories is analogous to the difference between doing geometry on a flat sheet of paper or a small patch of flat ground, and on the whole globe of the Earth, where parallel lines like the meridians of longitude *do* meet, and the angles in triangles can add up to more than 180 degrees.

* Newton's argument against collapsing universe: Smoot and Davidson, p. 27.
* Time slowing down for an astronaut: This is called the 'twin paradox', and the calculation showing how the travelling twin aged fourteen years while the Earth-bound twin aged fifty years is given in Schutz, pp. 28–30. Experimental tests of the twin paradox: Taylor and Wheeler, p. 133. For a detailed but 'popular' account of the twin paradox, and other implications of special relativity, see Davies (*About Time*), pp. 59*ff.*
* The theory and history of Einstein's cosmological term: Davies (*About Time*), pp. 135–40, pp. 159–60; 'clues' in the equations, p. 136. For technical details of Friedman's model (without cosmological term): Wald, pp. 99, 101. For a popular account of the use of the cosmological term to explain recent cosmological evidence that the expansion of the universe is accelerating, and for the possible consequences of this for the universe, see Lemonick, pp. 44–52.
* Doppler shift: see e.g. Smoot, p. 47; I have taken the idea that the waves are 'compressed' and 'stretched' from Davies (*About Time*), p. 88; Davies notes that the Doppler shift is used in police radar traps designed to catch speeding motorists.
* More on the Big Bang: see chapter 6 of Davies (*About Time*) for a detailed discussion of the complex issues involved in assessing the Big Bang theory. See also Hawking (especially for an alternative view on the beginning of time, e.g. p. 142; see also Davies, *ibid.*, pp. 188–92).
* The existence of the cosmic microwave background radiation was first predicted theoretically, using physical arguments and mathematical calculations, by the Russian-born George Gamow, working in America at Johns Hopkins University, and his colleagues Ralph Alpher and Robert Herman.
* On the inseparability of the cosmic background radiation and space: Smoot and Davidson, p. 85, and Gribbin, p. 17. On the expansion of the universe and the background radiation: Gribbin, pp. 219, 226, and Smoot and Davidson, pp. 85–6, 177.
* On gravitational accretion: Smoot, p. 187, 285, Gribbin pp. 37, 41, 60, Davies (*About Time*) pp. 147–8, 152–3, Penrose, p. 438; the power of 'subatomic'-scale gravity in the early 'micro'-universe: Gribbin, p. 161, 180–1, Weinberg, p. 135.

* The effect of gravity on the background radiation: Smoot, p. 285, Davies (*About Time*), p. 88 (gravitational red-shift), p. 152.
* Einstein on Newton, Faraday and Maxwell: Einstein (*Commemoration Volume*), pp. 66, 71.

MAXWELL'S MATHEMATICAL LANGUAGE
Setting the scene

* Maxwell and statistical mechanics: Planck, pp. 48–9; Maxwell's statistical equation and its verification (by Miller and Kusch in 1955): Resnick and Halliday, pp. 602–7. Quantum mechanical subtleties, see e.g. White and Gribbin, p. 182.
* Maxwell, molecules, Darwin, and design in nature: Hutchinson, p. 12.
* Faraday calling 'Ho Maxwell!': Goldman p. 92; also on this page of Goldman is the detail about Maxwell's house in Kensington, and his experiment with the 'coffin'; further detail on Mrs Maxwell's role in the gas experiments: J.J. Thomson, p. 16.
* Maxwell and colour photography: My description of his process is based on the entry on Maxwell in the *Dictionary of Scientific Biography*, and an online article by Laidler (which clarifies some common misconceptions about Maxwell's achievement, including the distinction between Maxwell's image of an everyday object and a previous coloured image of a light spectrum). See also *Chambers Dictionary of World History*, p. 643; *How Is It Done?* p. 235, and Segre, p. 159.
* Maxwell on returning to Glenlair to 'fraternize with the frogs': quoted in Goldman, p. 92. His duties as laird are from Tolstoy, p. 133.
* Goldman's commentary on the *Treatise*, Goldman, p. 159.
* Stephen Hawking's comment on his own 'unreadable' text: *A Brief History of Time*, p. vii.
* Maxwell on redeeming the character of mathematicians: Letter to H. R. Droop, Campbell, p. 291.
* Maxwell comparing and contrasting Ampère's and Faraday's scientific language: *Treatise*, pp. 175–6; for a proof (and discussion of the significance) of the particular result of Ampère's to which Maxwell was referring, see Resnick and Halliday, p. 853.
* Tait on Maxwell's mathematical ability: Knott, p. 262; Maxwell on his 'invincible stupidity': quoted in Goldman, p. 132.

Vector fields

* Maxwell on translating into vector language Faraday's concept of numbers of lines of force: *Treatise*, p. 189, article 541.
* On vectors: Goldman (p. 101) gives a good discussion of the conceptual significance of whole-vector notation, and Maxwell's approach to it. For historical controversy over vectors versus quaternions, see Boyer, p. 586, Struik, p. 172. This 'none too genteel' controversy occurred after Maxwell wrote his equations (Maxwell having been a pioneer in the field), and it justifies his decision to present his equations in component rather than vector/quaternion form.
* Hamilton's joyous letter to Tait (April 1859) reproduced in Knott.
* On Thomson and Tait's disagreement over vectors (or quaternions): Knott, p. 185.
* My description of Tait's study is taken directly from Knott; Tait on filling his pipe, p. 33; Thomson's recollection about Tait's charcoal list, p. 43.
* Maxwell on 'the electricity of kissing': Harman (ed.), Vol. 2, p. 766.
* 'O T', I am desolated': Goldman p. 159, footnote.
* On vector multiplication: There is another way of multiplying two vectors, called 'scalar multiplication'; it *is* commutative, but it does not have a direction – it does not produce a third vector, just a number determined by the two vectors' components.
* Maxwell's 'rough-hewn' names, and Knott's comments: Knott p. 167.
* Maxwell on the Newtonians' failure: 'the wrong philosophical perspective', *Treatise*, p. 158 (article 502), p. 492 (article 865), and 'the wrong mathematical language': *Treatise*, p. 176–7 (articles 528–9).

A holistic paradigm

* Maxwell's use of differential calculus over integral calculus: Maxwell's own words are in the *Treatise*, pp. 176–7; for further discussion of the significance of Maxwell's choice, see Goldman, pp. 142–3. Maxwell's reworking of Ampère's equation in differential form, see *Treatise*, p. 251. Note, Maxwell did also use some integration techniques for certain calculations, particularly Stokes's theorem. For Green's work on partial differential equations, see Struik, p. 169.
* Einstein praising Maxwell's use of partial differential equations, Einstein (*Commemoration Volume*), pp. 67–71.

* On the magnetic field vectors **E** and **B**: **E** is directly proportional to the electric force, but the relationship between **B** and magnetic force is more complicated. See *Treatise* pp. 233–4, 240 and Resnick & Halliday pp. 815, 816, 919.

* The **B** vectors 'wrapped around the current': Lodge, p. 125.

Maxwell's equations

* FitzGerald's comment: quoted in Hunt, p. 128.

* '... all the thought has been built into the symbolism': *The Illustrated Reference Book of Science*, p. 24.

* **On Maxwell's equations: Component versus vector form:** (see also the appendix in Hunt):

Maxwell's 'curl' equations: The curl **B** equation appears in the *Treatise*, in brief quaternion notation, on p. 258. It can be derived from the component equations (A) and (E), with the help of various combinations of equations (L), (H), (H*), (I*) and (F), pp. 233, 251–4.

The curl **E** equation was given by Maxwell in words in an 1868 paper (see Hunt, p. 126), but in the *Treatise* it is buried in a more detailed equation, given in component form as equations (B) on p. 239, and in vector notation (quaternion form) as equation (10) on p. 241. Unaware of Maxwell's 1868 paper, Maxwell's disciple Oliver Heaviside applied some grammatical rules to the *Treatise* equation (B)/(10) to produce, in 1884, the specific curl **E** equation in the modern form of Maxwell's equations. (See Hunt p. 127.)

Maxwell's 'divergence' equations: These were given in component form in the *Treatise*, p. 248 (article 604) and p. 254 (article 612); they follow from equation (A) (p. 233) for the div **B** case, and from equations (F) and (J) (pp. 252, 254) for div **E**, as Maxwell showed.

In other words, Maxwell's equations, as he wrote them, contained all that we now refer to as the four 'Maxwell's equations', and more. Maxwell says specifically, on p. 257 (article 618) of the *Treatise*, 'In this treatise we have endeavoured to avoid any process demanding from the reader a knowledge of the *Calculus* of [Vectors]. At the same time, we have not scrupled to use the *idea* of a vector when it was necessary to do so' (my italics). His caution was well advised, given the later controversy which surrounded quaternion and vector notation (see earlier note, 'On vectors', in the section headed **'Vector fields'**).

* 'Cassio' comment clarified by Harman.
* Maxwell on vector versus component forms: Knott, e.g. pp. 101, 151.
* On Euler and Newton's Second Law: Cushing, chapter 7. Note that the modern form F = ma did not actually appear in *Principia*. Newton used an equivalent form using momentum instead of acceleration.
* The modern form of Maxwell's equations given here are from Strauss, p. 339; different authors may use slightly different forms of these equations, depending on the units of measurement used; some authors, including Maxwell himself, use not only **E** and **B**, but also multiples of these vectors, for instance, Maxwell also uses **H**, where **B** = μ**H**, μ being a constant called the coefficient of magnetic permeability (see *Treatise*, equation (L), p. 254); by choosing different scales or units, some of these constants can be set to 1, so they do not appear explicitly in the equations.
* On Blake's poem: See Roger Jones, pp. 167–8, for a very different comparison between mathematics and Blake's lines.

MAXWELL'S RAINBOW
A bold mathematical prediction

* Faraday's law in words: see Maxwell, *Treatise*, p. 179, article 531, for Maxwell's exact (slightly more technical) wording.
* Maxwell's mathematical derivation of the electromagnetic 'wave equation', *Treatise*, pp. 434–6 (articles 784–6); here he obtained his wave equations not simply by further differentiation but by first *combining* and *rearranging* what are essentially his curl **B** and his curl **E** equations. The wave equation is a partial differential equation, so if Maxwell had not used this language in formulating his equations, he would not have discovered the wave equation.

  The wave equations are easier to obtain (by direct differentiation, as I mentioned in the text, although this process also needs the *combination* of the four different equations) from the modern vector forms of these equations. For such a derivation, see Strauss, pp. 340–1. Maxwell used direct differentiation in his 1868 paper, see Siegel, pp. 153–4.
* Waves as transmission of disturbances: see Resnick and Halliday, pp. 464–5, especially figure 19.1a, which I've adapted in my diagram.
* My diagram of the wave of changing electric force vectors is adapted

from Resnick and Halliday, p. 1147, and also from Maxwell's figure 67, *Treatise*, p. 439.

* Maxwell's prediction that light is electromagnetic (my emphasis), quoted in Torrance's Introduction (p. 16) to Maxwell's *A Dynamical Theory of the Electromagnetic Field*. For his explanation that light waves consist of waves (or disturbances) of magnetic and electric intensity, see *Treatise*, p. 439 (article 791), and also his popular lecture 'On Faraday's Lines of Force', 1873, Harman (ed.), Vol. 2, p. 810. (In this lecture he also uses the bell-wire analogy of his childhood to describe a wave propagation, p. 795, and speaks of the field advocates being in the Newtonians' 'enemy's camp', p. 796).

On light as waves of electric and magnetic force, also see J.J. Thomson, p. 39, and see Maxwell's discussion of the wave-like 'radiation of light and heat as well as the forces of electricity and magnetism', in his paper *On Electromagnetism, 1872–3*, Harman (ed.), Vol. 2, pp. 772–3.

* 'To few men in the world has such an experience been vouchsafed': Einstein, *Ideas and Opinions*, p. 327; Maxwell's 'great guns', quoted in Torrance's Preface (p. ix) to Maxwell's *A Dynamical Theory* ...

* On Hertz's method of detecting electric waves: My description is adapted from Murchie, pp. 443–4 (Hertz's quote is from this reference, p. 444), and also from *How Is It Done?* p. 217, and J.J. Thomson, pp. 41–3. See also Lodge's account of his own discoveries, and his inspiration from Maxwell, Lodge, e.g. p. 128. For an account of Hertz's process, involving a series of experiments from 1886 to 1888, see Rowlands, pp. 23–26.

Ripples through history

* Carrying sound on radio waves: *How Is It Done?* pp. 215–17. Fessenden's broadcast could be heard 160 kilometres away, p. 215.

* On the loss of energy of waves due to radiation: Lodge, p. 127; Resnick and Halliday, pp. 478, 1077–8.

* Maxwell's field model of gravity (in response to an idea of Faraday's): Letter to Michael Faraday, Harman (ed.), Vol. 1, p. 550; this is the particular letter that occasioned Faraday's enthusiastic response as mentioned in an earlier chapter. See also Maxwell's *Lecture on Faraday's Lines of Force*, 1873, Harman (ed.), Vol. 2, pp. 810–11.

* On the electromagnetic nature of matter, and the production of

magnetism and static electricity, see e.g. *The New Book of Knowledge*, Vol. 12 (M), p. 28, and Vol 5 (E), pp. 123–4; also Tolstoy, p. 109.

Note that Ampère had postulated the idea that magnetism was created by microscopic electric currents, although he had no idea of how this actually happened. See Maxwell's *Treatise*, pp. 472–4 (articles 833–7). See also Resnick and Halliday, footnote, p. 936.

* On the history of radio: *How Is It Done?* p. 215; White; Neville; Howeth; Loomis.

A 'war' over vectors

* Boyer on Maxwell: p. 557.
* Thomson on Pecksniff, and Tait's 'Pecksniffian' ways: Knott
* Thomson and the Atlantic cable: Segre pp. 208–9; Maxwell's letter: Hartman, Vol. 1, p. 555.
* An 'epidemic of telegraphic engineers': Wilson

IMAGINING THE WORLD WITH THE LANGUAGE OF MATHEMATICS: A REVOLUTION IN PHYSICS

* Contemporary reactions to the *Treatise* by Hertz, Poincare and Thomson: quoted by Goldman, pp. 158–9; Poincare later introducing the theory to France, Segre, p. 166. Hertz's use of the equations only is discussed by Buchwald, pp. 190–93, and Hertz's acceptance of the equations as primary, Siegel, p. 167. Thomson on Maxwell's mysticism, quoted in Torrance's Preface (p. ix) to Maxwell's *A Dynamical Theory* ...

A methodological paradigm-shift: The slow transition from concrete models to mathematical imagination

* Maxwell's caution regarding light waves: Jeans, p. 99.
* 'In space, no-one can hear you scream': I remember this line and I believe the movie was *Alien* (see Krauss).
* The ether idea – and the light wave hypothesis – had been around since Newton's day, but it took on new vigour and complexity in the nineteenth century when Young's wave theory was accepted.
* The magneto-optical effect being the 'talk of the town' after it was announced at 'what we may consider the world's first science press conference', Agassi, p. 295.
* Maxwell's conclusion that light and electromagnetism shared the same

ether, *Scientific Papers*, quoted Torrance, p. 14, introduction to Maxwell's *A Dynamical Theory* ... Maxwell told Thomson how surprised he was when his ether wave speed matched the measured speed of light: Harman (ed.), Volume 1, p. 695.

* On Maxwell's philosophical rigour: The significance of Maxwell's approach is powerfully expressed by Goldman, pp. 134, 155.

* Maxwell on unverified assumptions: *Treatise*, Vol 2, p. 218.

* Maxwell on his 'mathesis', quoted by Torrance, Introduction to Maxwell's *A Dynamical Theory* ..., p. 5.

* Hertz on the wisdom of mathematical equations: Schwartz and McGuinness, p. 134.

* Maxwell's belief in the 'probable' existence of the ether was temperate compared even with that of some of his later disciples. See Rowlands, pp. 9, 11: both FitzGerald and Lodge believed in the ether, partly for mystical reasons, 'perhaps influenced by Tait and Balfour Stewart's religiously inspired *Unseen Universe*, (1875)'.

But see Goldman for an account of Maxwell's confused thinking on the ether implications of his equations, pp. 202–4.

* The orbital speed of the Earth: Davies (*About Time*), p. 127.

* Einstein riding on a light wave: Schwartz and McGuinness, pp. 70*ff.*

* See Resnick, p. 25, for details of the various subsequent versions of the Michelson–Morley experiment.

* On 'c', the speed of light: 'c' has now become the standard symbol for the speed of light, but Maxwell had actually used 'V' (for velocity).

* Some physicists are still critical of the constant speed of light idea, and better experiments (e.g. Marinov) may yet find evidence of an absolute frame of reference; but the ether is no longer seen as a necessary mechanical device for the transmission of light and gravity.

* Einstein on the need for a more dynamic view of space: 'The Problem of Space, Ether and the Field in Physics', *Ideas and Opinions*, pp. 276–85, esp. p. 281; see also p. 345.

The Maxwellians

* For use of the term 'Maxwellians', see Rowlands, p. 15, and Hunt. The story about Lodge buying the *Treatise* is from Hunt, p. 26. Lodge's origins, and his inspiration from the 1873 meeting at which Maxwell's

Treatise was discussed, including enrolling 'at the relatively late age of 22', as a full-time student, Rowlands, p. 9.

For details of the presidential address on the *Treatise*, see Lodge, p. 126, and see p. 128 for his production of waves. See p. 127 for his meeting with FitzGerald, in 1878, and their subsequent collaboration. For more details on Lodge's demonstration of waves, see Rowlands, pp. 19–22.

* Lodge on Hertz, quoted in Rowlands, pp. 32–26. See p. 31 for discussion of the role of the British Maxwellians in clarifying and promoting Hertz's discovery.

* On the *Treatise* being praised by a pure mathematician rather than a physicist: On the other hand, Hunt presents a case in which it was pure mathematicians (as well as practical engineers) who gave Heaviside a hard time (just as Newtonian mathematicians like Airy had given Faraday a hard time): Hunt (in Dear), pp. 72–95.

* Heaviside on Maxwell as a bright star: quoted by Hunt, p. 4.

* Faraday's field ideas and telegraphy: Hunt, pp. 64–5; Heaviside taught himself enough maths to extend Thomson: Hunt, pp. 64–5.

* For my description of how a telephone works, I am indebted to *How Is It Done?* pp. 214, 225.

* Challenges to Bell's claim to the invention of the telephone: There is much information online, e.g. Meucci, see e.g. http://www.Popular-Science.net/history/meucci_resolution.html.

For other challenges, see Evenson (on McDonough); Bellis (on Gray as co-inventor with Bell).

* Maxwell's telephone demonstration: Goldman, p. 188. Heaviside's telephone improvements: Hunt, p. 129, 137. Heaviside's fight with the 'practical men': Hunt, p. 129. Resistance to Thomson's theoretical ideas: Resnick and Halliday, p. 1.

* Heaviside's enthusiasm for science, his philanthropic motives etc: quoted by Hunt, p. 57; his seeking the laws of God: Hunt, p. 52; his self-publishing venture: Hunt, p. 145.

* On Maxwell's death: His last letter to Tait: Goldman p. 194 – Goldman has 'the essentially finite variety of the sensations …', and I have supposed this to be a misprint (perhaps even on Maxwell's part) and have replaced it with 'infinite'; 'Christian gentleman' quoted by Tolstoy, p. 168; his specialist's comments on his 'calmness of mind',

quoted by Tolstoy, p. 169; his last wishes to his doctor: Tolstoy p. 169; McKenzie's account of Maxwell's last moments: Tolstoy, p. 170; his grave: Tolstoy, p. 171; Campbell's last visit to Glenlair: Goldman, p. 195.

* Tait hated emotionalism, and did not leave Edinburgh (or St Andrews) after the mid 1870s: Knott.

* Heaviside's concern about Thomson, and his recognition of FitzGerald as an ally: Hunt, p. 145.

* Thomson's class sizes dwindling because he didn't teach Maxwell's theory: Wilson, p. 66; Rutherford's letter, *ibid.*, p. 216; Thomson saying he knew no more about electricity and magnetism in 1895 than he did 50 years earlier: *ibid.*, pp. 171–2, and also Young, p. 35.

* Thomson not forgiving Maxwell: Siegel, p. 160.

* Max Planck on Maxwell: Planck, pp. 57–8.

* James Clerk Maxwell Foundation, see 'A visit to James Clerk Maxwell's house' by O'Connor and Robertson. This article is also my source for the statement by Einstein, 'One scientific epoch ended and another began with James Clerk Maxwell'.

EPILOGUE

* Einstein's quote re relativity and Maxwell's theory is from his paper *On the Electrodynamics of Moving Bodies*, quoted in Schwartz and McGuinness, p. 95 (my emphasis). See also Resnick, p. 35.

* On quantum physics: Hoffmann, p. 192; Strauss, pp. 354–5; Bethe and Jackiw, pp. 350–1; D'Inverno, pp. 49–50; Resnick and Halliday, p. 1183; Davies and Gribbin, pp. 230–3; Penrose, chapter 6, and see pp. 372–3 for Maxwell's and Dirac's equations as the basis of QED.

Planck had proposed the idea of light quanta as a mathematical technique to explain certain heat radiation anomalies (the problem of 'black body radiation'); Einstein proved this idea in his analysis of the photoelectric effect. For a discussion of this, and also of the role of mathematics in guiding De Broglie's idea of matter waves, see White and Gribbin, pp. 85–92, 178–9. For the idea of light and matter – specifically, photons and electrons – behaving as both particles and waves at the same time, see White and Gribbin pp. 180–82. For the automatic bank teller analogy of a quantum, see White and Gribbin p. 87. For a different, really neat approach to Planck's quantum hypothesis for black-body radiation, see Hawking, p. 58, and see pp. 57–66 for an excellent discussion of quantum theory.

* The importance of mathematical wave mechanics for our understanding of atoms and the sciences and technologies based on them: Hawking, pp. 60, 63–4. (For the role of QED in radiobiology, there are many references on the web, e.g. Google search 'quantum electrodynamics' or 'synchrotron biology'.)

* Note that QED applies on the scale of the electron; a different theory might be needed at even smaller scales.

* Dirac's disbelief in his mathematical positrons – and his later admission that his equations were smarter than he – and Rutherford's comment: Crease and Mann, pp. 60–79 (pp. 78–9 for the Dirac and Rutherford quotes, p. 75 for Weyl, see also p. 68 for Dirac's use of Maxwell's field theory). Also see Ferris, pp. 80–4, for Dirac's own account of his mathematical discovery of positrons, taken from his 1933 Nobel Prize address. On the experimental discovery of positrons, see Weinberg, p. 76, and Davies (*About Time*), p. 204.

* Maxwell's equations apply (with QED) accurately over an extraordinary scale, from the subatomic to the galactic: Penrose, p. 198; see also Cropper. For the accuracy of the measurement of an electron's magnetic strength, and for Feynman's analogy of the distance between NY and LA, see Penrose, p. 199.

* Mathematics and liberty: Pauwels and Bergier, e.g. p. 242 (quoting Cantor).

* Maxwell's equations were not taught unencumbered by physical models and assumptions till about 1930: Bell, p. 127.

* 'Agonizing appeals to a nonexistent physical intuition': Bell, p. 128; Dirac's quote, Bell, p. 128.

* On mathematics and consciousness: For a fascinating discussion of mathematics and universal consciousness, see Davies, *Are We Alone?*, pp. 82–5, refer also to pp. 57–8. See also Roger Jones, for a discussion of the relationship between the 'metaphors' of physics and physical reality, and the role of consciousness in 'creating' the universe (e.g. p. 5).

BIBLIOGRAPHY

Joseph Agassi, *Faraday as a Natural Philosopher,* University of Chicago Press, 1971

Aristotle, *Categories,* in Steven M. Cahn (ed.)

Aristotle, *Physics (Book II)),* in Steven M. Cahn (ed.)

Henri Becquerel, 'Déviation du rayonnement du radium dan un champ électrique', *Comptes Rendus,* 26 Mars, 1900, 809-15

Eric Temple Bell, *Mathematics, Queen and Servant of Science,* Tempus Books (Microsoft Press and the Mathematical Association of America), 1987 (original publication, 1951)

Mary Bellis, *The History of the Telephone: Alexander Graham Bell, Elisha Gray, and the Invention of the Telephone,* available online at http://inventors.about.com/library/inventors/telephone.htm

Hans Bethe and Roman Jackiw, *Intermediate Quantum Mechanics,* W.A. Benjamin Inc, 1973

David Bodanis, *E = mc²: A Biography of the World's Most Famous Equation,* Pan, 2000

Carl Boyer, *The History of Calculus and Its Conceptual Development,* Dover, 1959. (This is referenced in the 'Sources' as 'Boyer (1959)'. The next-listed book is the reference when 'Boyer' alone is cited.)

Carl Boyer, (revised by Uta Merzbach), *A History of Mathematics,* John Wiley and Sons, 1991

Jed Z. Buchwald, *From Maxwell to Microphysics,* University of Chicago Press, 1985

Dionys Burger (translated by Cornelie Rheinboldt), *Sphereland: A Fantasy about Curved Spaces and an Expanded Universe,* Thomas Y. Crowell Co., NY, Apollo Edition, 1968

James Burke, PBS series 'Connections', quoted in an online article 'Sir Isaac Newton', available online at http://cns.pds.pvt.k12.ny.us/~jonathan/newton.html

Steven M. Cahn, *Classics of Western Philosophy*, Hackett Publishing Co., Indianapolis/Cambridge, 1990

Malcolm Cameron, *Heritage Mathematics*, Eward Arnold Australia, 1984

Lewis Campbell and William Garnett, *The Life of James Clerk Maxwell*, first published in 1882. My reference is to a reprint of this edition, with a preface and appendix by Robert H. Kargon, Johnson Reprint Corporation, NY, 1969. (In 1997, James Rautio of Sonnet Software made the original 1882 version available online at http://www.sonnetusa.com/bio/maxwell.asp)

Geoffrey N. Cantor, 'Reading the Book of Nature: The Relation between Faraday's Religion and his Science', in David Gooding and Frank A.J.L James (eds)

Rory Carroll, 'Bell did not invent telephone, US rules', in *The Guardian*, 17 June 2002.

Robert Crease and Charles Mann, 'The Man Who Listened', in Ferris (ed.)

William Cropper, *Great Physicists*, Oxford University Press, 2001

James T. Cushing, *Philosophical Concepts in Physics*, Cambridge University Press, 1998

Paul Davies and John Gribbin, *The Matter Myth*, Viking, 1991

Paul Davies, *The Mind of God*, Simon and Schuster, 1992

Paul Davies, *About Time*, Viking, 1995

Paul Davies, *Are We Alone? Implications of the Discovery of Extra-terrestrial Life*, Penguin, 1995

Peter Dear (ed.), *The Literary Structure of Scientific Argument*, University of Pennsylvania Press, Philadelphia, 1991

Rene Descartes, *Meditations on First Philosophy*, in Steven M. Cahn (ed.) *Dictionary of Scientific Biography* (editor-in-chief, Charles Coulston Gillispie), Scribner, 1970–1990

Ray D'Inverno, *Introducing Einstein's Relativity*, Clarendon Press, Oxford, 1993

Paul Dirac, 'Theory of Electrons and Positrons', (Nobel Prize address, 1933), in T. Ferris (ed.)

Albert Einstein, 1907 article on mass-energy, see H.M. Schwartz, part II, pp 813-14, for translation

Albert Einstein, *Ideas and Opinions*, Three Rivers Press, NY (copyright 1982, Crown Publishers). (Based on *Mein Weltbild*, edited by Carl Seelig and other sources. New translations and revisions by Sonia Bargmann)

Albert Einstein, 'An elementary derivation of the equivalence of mass and energy', in A. P. French (ed.), (originally published in *Out of My Later Years*)

Albert Einstein, 'The Problem of Space, Ether and the Field in Physics', in *Ideas and Opinions*

Albert Einstein, 'Geometry and Experience', quoted in A.P. French (ed.)

Albert Einstein, 'Maxwell's Influence on the Development of the Conception of Physical Reality', in *James Clerk Maxwell: A Commemoration Volume, 1831–1931,* Cambridge University Press, 1931

Evar Ekeland, *Mathematics and the Unexpected,* University of Chicago Press, 1988

A. Edward Evenson, 'The McDonough Transmitter', available online at http://atcaonline.com/phone/McDonough.html

Michael Faraday, *Experimental Researches in Electricity,* Volumes 1, 2, 3, Dover, 1965. (Volumes 1 and 3 were originally published by Richard and John Taylor, 1844.)

J. Fauvel and J. Gray, *The History of Mathematics: A Reader,* Macmillan Education Ltd, 1987

T. Ferris (ed.), *The World Treasury of Physics, Astronomy and Mathematics,* Little, Brown and Company, 1991

Sophie Forgan, 'The Institutional Context', in David Gooding and Frank A.J.L. James (eds.)

Grant R. Fowles, *Analytical Mechanics,* Holt, Rinehart and Winston, 1962

A. P. French (ed.), *Einstein: A Centenary Volume,* Heinemann, 1979

Martin Goldman, *The Demon in the Aether,* Paul Harris Publishing, Edinburgh, 1983

David Gooding and Frank A.J.L. James (eds.), *Faraday Rediscovered,* Macmillan, UK, 1985

John Gribbin, *In the Beginning,* Viking, 1993

Michael Guillen, *Five Equations That Changed the World,* Abacus, 1999

P.M. Harman (ed.), *The Scientific Letters and Papers of James Clerk Maxwell,* Vol 1 (1846–1862), Vol 2 (1862–1873), Cambridge University Press, 1990 (Vol 1), 1995 (Vol 2)

P.M. Harman, *The Natural Philosophy of James Clerk Maxwell,* Cambridge University Press, 1998

Robert A. Hatch, *Sir Isaac Newton,* available online at http://www.class.ufl.edu/users/rhatch/pages/01–Courses/current-courses/08sr-newton.htm

Stephen Hawking, *A Brief History of Time,* Bantam, 1991

Hal Hellman, *Great Feuds in Science,* Joseph Wiley and Sons, 1998

David Hochfelder, 'Joseph Henry: Inventor of the Telegraph?', in *The Joseph Henry Papers Project* of the Smithsonian Institute, 1998, available online at http://www.si.edu/archives/ihd/jhp/joseph20.htm

Banesh Hoffmann (with Helen Dukas), *Einstein,* Paladin, 1977

Captain Linwood S. Howeth, *History of Communications – Electronics in the United States Navy,* 1963, available online at http://www.angelfire.com/nc2/whitetho/1963hw02.htm

How Is It Done? Reader's Digest, 1995

Bruce Hunt, 'Rigorous Discipline: Oliver Heaviside versus the Mathematicians', in Peter Dear (ed.)

Bruce J. Hunt, *The Maxwellians,* Cornell University Press, 1991. (This is the reference I mean when citing 'Hunt' in the 'Notes and Sources'. Otherwise I cite 'Hunt, in Dear'.)

Ian Hutchinson, *James Clerk Maxwell and the Christian Proposition*, MIT IAP Seminar, 'The Faith of Great Scientists', 1998, available online at http://silas.psfc.mit.edu/Maxwell/maxwell.html

Herbert E. Ives, 'Derivation of the Mass-Energy Relation', *J. Optical Soc. Am.*, 42, pp 540-43

Frank James (ed.), *The Correspondence of Michael Faraday, Vol 2*, Institution of Electrical Engineers, UK, 1993

James Jeans, 'Clerk Maxwell's Method', in *James Clerk Maxwell: A Commemoration Volume, 1831–1931*, Cambridge University Press, 1931

Roger Jones, *Physics as Metaphor*, Meridian, 1982

R. Victor Jones, web-page on the history of telegraphy at htt://people.deas.harvard.edu/~jones/oscie129/lectures/lecture5/elecmag_tel/morse_tel.html

Debra Jopson, 'Speaking Other Universes', *The Age*, Saturday Extra, 3 March, 2001

Morris Kline, *Mathematics: The Loss of Certainty*, Oxford University Press, 1980

David Knight, 'Davy and Faraday', in David Gooding and Frank A.J.L. James (eds.)

C.G. Knott, *The Life and Scientific Work of P.G. Tait*, Cambridge University Press, 1911

Lawrence Krauss, *The Physics of Star Trek*, Basic Books (Harper Collins), 1995

Keith J. Laidler, 'The Genius of James Clerk Maxwell, Part 2', published on the University of Waterloo's website, www.science.waterloo.ca/physics/p13news/maxall.html

Ian Lancashire (ed.), 'Selected Poetry of James Clerk Maxwell (1831–1879)', in *Representative Poetry Online*, Department of English, University of Toronto, 2003, http://eir.library.utoronto.ca/rpo/display/poem2707.html

Gottfried Leibniz, *On Newton's Mathematical Principles of Philosophy (Letter to Samuel Clarke)* in *Leibniz Selections*, ed. Philip Wiener, Charles Scribner & Sons, NY, 1951

Michael Lemonick, 'The End', *Time*, 25 June, 2001

Bruce P. Lenman (consultant ed.), *Chambers Dictionary of World History*, 2000

Oliver Lodge, 'Clerk Maxwell and Wireless Telegraphy', in *James Clerk Maxwell: A Commemoration Volume, 1831–1931,* Cambridge University Press, 1931

Mahlon Loomis, lecture, published in *Radio News*, November 1922, available online at http://angelfire.com/nc/whitetho/1872loom.htm

B. Lynch and R. Parr, *Maths 10*, Longman Sorrett, 1982

Magnus Magnusson (ed.), *Chambers Biographical Dictionary*, 1996

David Malouf, 'The only speaker of his tongue' in 'Antipodes', Chatto and Windus/The Hogarth Press, 1985

David Malouf, *Remembering Babylon*, The Softback Preview, by arrangement with Chatto and Windus, 1993

Stefan Marinov, 'Measurement of the Laboratory's Absolute Velocity', *Gen. Rel. Gravit.* 12 (1980), p 57ff

James Clerk Maxwell, *Treatise on Electricity and Magnetism*, Volumes 1 and 2, Dover, NY, 1954 (reprint of the 1891 third edition by Clarendon Press)

James Clerk Maxwell, *Lecture on Faraday's Lines of Force*, in Harman, Volume 2

James Clerk Maxwell, inaugural lecture at Aberdeen, Marischal College, 3 November 1856, in Harman, Vol 1

James Clerk Maxwell, *A Dynamical Theory of the Electromagnetic Field*, introduced by Thomas Torrance, Scottish Academic Press (Edinburgh), 1982

Daniel P. McVeigh, 'An Early History of the Telephone, 1664–1865', 2000, ILT, Columbia, available online at http://www.ilt.columbia.edu/projects/bluetelephone/html/

Mario Meurier (ed.), *Femmes Pythagoriciennes: Fragments et lettres de Theano, Perictione, Phintys, Melissa et Myia*, traduction nouvelle avec prolegomenes et notes par Mario Meurier, L'Artisan du livre, Paris, 1932

Henrietta Midonick (ed.) *The Treasury of Mathematics I*, Pelican, 1968

Yoshio Mikami, 'The Development of Mathematics in China and Japan', in Henrietta Midonick (ed.)

Guy Murchie, *Music of the Spheres, Vol 2*, Dover, NY, 1967

Cherise Neville, *Greenbelt Museum Artifact Study*, online at http://www.otal.umd.edu/~vg/amst205.S97/vj19/Project4.html

J.J. O'Connor and E.F. Robertson, *Sir Isaac Newton*, available online at www.groups.dcs.st-and.ac.uk/history/Mathematicians/Newton.html

J.J. O'Connor and E.F. Robertson, 'A visit to James Clerk Maxwell's house', November 1997, available online at www-history.mcs.st-andrews.ac.uk/history/HistTopics/Maxwell_House.html. (The house is owned by the James Clerk Maxwell Foundation)

Oxford Reference Encyclopedia, Oxford University Press, 1998

Louis Pauwels and Jacques Bergier, *Morning of the Magicians*, (translated from the French by Rollo Myers), Stein and Day, 1964

Roger Penrose, *The Emperor's New Mind*, Vintage, 1989

Max Planck, 'Maxwell's influence on Theoretical Physics in Germany', in *James Clerk Maxwell: A Commemoration Volume, 1831–1931*, Cambridge University Press, 1931

William Rankin, *Newton for Beginners*, Allen and Unwin, 1993

R. Resnick and D. Halliday, *Physics*, John Wiley and Sons, 1966; (a new extended edition, *Fundamentals of Physics, Extended*, by D. Halliday, R. Resnick and J. Walker, is now available from Wiley)

Robert Resnick, *Introduction to Special Relativity*, John Wiley and Sons, 1968

H.L. Resnikoff and R.O Wells Jr, *Mathematics in Civilization*, Dover, NY, 1973

Peter Rowlands, *Oliver Lodge and the Liverpool Physical Society*, Liverpool University Press, 1990

A. Roy and D. Clarke, *Astronomy: Structure of the Universe*, Adam Hilger, Bristol, 1982

Bernard Schutz, *A First Course in General Relativity*, Cambridge University Press, 1985

H.M. Schwartz, 'Einstein's comprehensive 1907 essay on relativity', parts I, II, III, *Am. J. Phys.* 45 (1977), 512-517, 811-817, 899-902

Joseph Schwartz and Michael McGuinness, *Einstein for Beginners*, Allen and Unwin, 1992

Emilio Segre, *From Falling Bodies to Radio Waves*, W.H. Freeman and Co., NY, 1984

Daniel Siegel, *Innovation in Maxwell's Electromagnetic Theory*, Cambridge University Press, Cambridge, 1991

George Smoot and Keay Davidson, *Wrinkles in Time*, Little, Brown and Co., 1993

Dava Sobel, *Galileo's Daughter*, Fourth Estate, London, 1999

Walter Strauss, *Partial Differential Equations*, John Wiley and Sons, 1992

Dirk Struik, *A Concise History of Mathematics*, Dover, 1967

Roger Sworder, 'Gravity's Harmony', in *Quadrant*, July–August, 1999, pp. 70–1

Edwin Taylor and John Archibald Wheeler, *Spacetime Physics*, W.H. Freeman and Co. NY, 1992

The Illustrated Reference Book of Science, Colporteur Press, Sydney, 1982

The New Book of Knowledge, Volumes 1–21, Grolier, 1982

D'Arcy Wentworth Thompson, *On Growth and Form*, first published in 1917; see abridged edition, edited by John Tyler Bonner, with a foreword by Stephen Jay Gould, Cambridge University Press, 1992

J. J. Thomson, 'James Clerk Maxwell', in *James Clerk Maxwell: A Commemoration Volume, 1831–1931,* Cambridge University Press, 1931

Ivan Tolstoy, *James Clerk Maxwell*, Canongate, Edinburgh, 1981

Voltaire, *Candide*, Classiques illustres Hachette, Paris (with expanatory notes and questions by Claude Blum)

Voltaire, *Lettres Philosophiques* (ed. Rene Pomeau), Garnier-Flammarion, Paris, 1964

Carol Vorderman, *How Maths Works*, Covent Garden Books (Dorling Kindersley copyright and original publishers), 1998

Robert Wald, *General Relativity*, University of Chicago Press, 1984

Steven Weinberg, *The First Three Minutes*, Bantam, NY, 1980

Margaret Wertheim, *Pythagoras' Trousers: God, Physics and the Gender Wars*, Fourth Estate, London, 1997

Richard Westfall, *Never at Rest: A Biography of Isaac Newton*, Cambridge University Press, 1980

Michael White and John Gribbin, *Einstein: A Life in Science*, Simon and Schuster, 1993

Thomas H. White, *United States Early Radio History*, available online at http://www.EarlyRadioHistory.us/index.html

Eugene Wigner, 'The Unreasonable Effectiveness of Mathematics in the Natural Sciences', in Ferris (ed.), pp. 534–5; originally published in 1960 by John Wiley and Sons

L. Pearce Williams (ed.), *The Selected Correspondence of Michael Faraday*, Vol 2, Cambridge University Press, 1971

David Wilson, *Kelvin and Stokes*, Adam Hilger, 1987

Sue Woolfe, *Leaning Towards Infinity*, Vintage, 1996

Benjamin Woolley, *The Queen's Conjuror*, Harper Collins, London, 2001

A.P. Young, *Lord Kelvin: Physicist, Mathematician, Engineer*, The British Council (published for the Council by Longmans, Green and Co., 1948)

INDEX